Light

Also by Bruce Watson

Freedom Summer: The Savage Season of 1964 That Made Mississippi Burn and Made America a Democracy

Sacco and Vanzetti: The Men, the Murders, and the Judgment of Mankind

Bread and Roses: Mills, Migrants, and the Struggle for the American Dream

The Man Who Changed How Boys and Toys Were Made

Light

A Radiant History from Creation to the Quantum Age

BRUCE WATSON

BLOOMSBURY
NEW YORK • LONDON • OXFORD • NEW DELHI • SYDNEY

Bloomsbury USA
An imprint of Bloomsbury Publishing Plc

1385 Broadway
New York
NY 10018
USA

50 Bedford Square
London
WC1B 3DP
UK

www.bloomsbury.com

First published 2016

ISBN: HB: 978-1-62040-559-8
ePub: 978-1-62040-561-1

LIBRARY OF CONGRESS CATALOGING-IN-PUBLICATION DATA HAS BEEN APPLIED FOR.

2 4 6 8 10 9 7 5 3 1

Typeset by RefineCatch, Limited, Bungay, Suffolk
Printed and bound in USA by Berryville Graphics Inc., Berryville, Virginia

For my mother, whose interest in everything
turned out to be the greatest gift

Contents

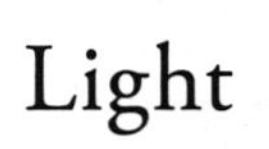

Light

Introduction

We eat light, drink it in through our skins.

—JAMES TURRELL, LIGHT AND SPACE ARTIST

Galileo was bewildered. Toward the end of his life, a life that had witnessed wondrous light that none had seen before, the great scientist confessed one failure. Decades had passed since a friend had given him several of the stones Italians called "solar sponges." Soaking up sunlight, emitting a soft green glow, the stones convinced Galileo that Aristotle had been wrong about light. It was not a warm, ethereal element. Light could be as cold as the moon and as corporeal as water. But what was it?

Over the years, Galileo had learned to reflect light, to bend it, to amaze observers with telescopes that spotted ships two hours' sail from Venice. Turning his telescope toward the night sky, he had been the first to see the moons of Jupiter and identify craters on the moon. Later he proposed the first experiment to clock the speed of light, bouncing lantern beams across the hilltops of Tuscany. Galileo never conducted his speed test. Other experiments, other trials, demanded his attention, yet he continued to wonder about light. Shortly before his death, blind and broken, he admitted how he longed for an answer. Though under house arrest for heresy, he said he would gladly suffer a harsher imprisonment. He would live in a cell with nothing but bread and water if, upon emerging, he could know the truth about light.

The truth is that, despite three millennia of investigation by humanity's most brilliant detectives, light refuses to surrender all its secrets. As familiar as our own faces, light is the first thing we see at birth, the last before dying. Some, having seen a warm glow as they flirted with death, swear that light will welcome us to another life. "Painting is light," the Italian master Caravaggio noted, and each day light paints a mural that sweeps around the globe, propelling us into the morning. Ever since the Big Bang, light has been stealing the show. And for countless scientists, philosophers, poets, painters, mystics, and anyone who ever stood in awe of a sunrise, light *is* the show.

"If there is magic on this planet," the naturalist Loren Eiseley wrote, "it is contained in water." Light, however, is the magician of the cosmos. Light makes darkness vanish and worlds reappear. Light opens each day with a blaring overture, then throws its wands to earth and casts diamonds on lakes and oceans. Each night, light's tricks make the stars seem alive. Seen through telescopes Galileo could never have imagined, light dances across the rings of Saturn, shapes gas clouds into crabs and horse heads, spirals from great galaxies and bursts from newborn stars. As reliable and relentless as time, light will begin tomorrow with another hurrah, then close the show with house lights, low and glimmering. Drawn to it as surely as any moth, we cannot live without it. "When the great night comes, everything takes on a note of deep dejection," the psychologist Carl Jung wrote, "and every soul is seized by an inexpressible longing for light."

But what is light? What meaning have our brilliant detectives found in it? Is it God? Truth? Mere energy? Since the dawn of curiosity, these questions have been at the core of human existence. The struggle for answers has given light a history of its own. In human consciousness, light first appeared in stories of creation, stories spun in the glow of firelight or torch. From the immortal "Let there be light" of Genesis to the Icelandic *edda* that had God throwing embers into the darkness, light is the primal ingredient of every creation story. Following its mythical creation, light matured into a mystery that intrigued philosophers from Greece to China. Was light atoms or shimmering eidola? Were we all, as the Apostle Paul wrote, "children of light"? Just when each sage had light pinned down, the mystery rose again, posing further questions, fresher metaphors.

Light was Jesus ("I am the light of the world"). No, it was Allah—"the Light of the heavens and the earth." No, it was the Buddha, also called Buddha of Boundless Light, Buddha of Unimpeded Light, Buddha of Unopposed Light . . . Light was inspiration—inner light. Light was love ("the light in her eyes"). Light was sex—divine coupling, according to Tantric Buddhism, fills sexual organs with radiance. Light was hope, thought, salvation ("seeing the light"). Dante filled his *Paradiso* with "the heaven of pure light." Shakespeare toyed with it: "Light, seeking light, doth light of light beguile." The blind poet Milton was obsessed with light: "Hail, holy light, offspring of Heav'n first-born." Caravaggio and Rembrandt captured light as a sword cutting through the blackness. Vermeer sent it streaming through windows. Beethoven heard it as French horns. Haydn preferred an orchestra's full blare. Meanwhile, ordinary people spoke of the light of freedom, the light of day, the light of reason, the light of their lives . . .

Yet from the first theories about its origin, light sparked bitter disputes. The earliest philosophers quarreled about light—was it emitted by the eye or by every object? Holy men debated whether light was God incarnate or merely His messenger. And then there was light's handmaiden—color. Was it innate in each object or merely perceived by the eye? While some debated, others celebrated in festivals of light ranging from Hanukkah and winter solstice to the Hindu Diwali and the Zoroastrian Nowruz. Meanwhile, without the slightest concern for science or religion, sunrise and sunset circled the planet, rarely disappointing those who paused to admire.

Traced from creation to the quantum age, light's trajectory suggests that the miracle has lost some luster. Once we spoke of "the light fantastic," but now we find light cheap and easy, available not just in every home and office but in every palm and pocket. Artificial light once came from precious few sources—candles, lamps, and torches. But with lighting now a $100 billion industry, beams shoot out of bike helmets, key chains, shower heads, e-readers, smartphones, tablets, and dozens more devices. Light has become our most versatile tool, healing detached retinas, reading bar codes, playing DVDs. Light from liquid crystal displays brings us the World Wide Web. Light through fiber-optic cables carries the messages that girdle the planet. Thus we have made light as common as breath, as precise as a laptop.

But there was a time when light waged a heroic battle with darkness. It was a time when night skies were not bleached by urban glare, when candles were not romantic novelties, when light was the source of all warmth and safety. For the vast majority of human history, each sunrise was a celebration; each waxing moon stirred hope of nights less terrifying. And to anyone caught unprepared—in dark woods, on echoing streets, even at home when lamps flickered and failed—light was, simply, life. Unlocking its secrets required an uncommon set of keys. Curiosity. Persistence. Mirrors, prisms, and lenses. Through the centuries, as civilizations took turns asking and answering, the keys passed from Greece to China to Baghdad, from medieval

France to Italy and back. When the keys came to Isaac Newton, his answers spread to the world, unlocking secrets we are still exploring.

Our evolving concepts of light chart the development of human thought, from spiritual to secular, from superstitious to scientific. So long as light was God, divinity incarnate arose each morning. Millennia passed with only a handful of curious men considering light less than holy. Then, during the 1600s, the Scientific Revolution gave curiosity the upper hand. Once Kepler saw light as subject to physical laws, once Galileo showed how to gather it, once Newton broke it into color, light was no longer merely God's essence. The rhapsody was finished. Enchantment had met its midnight.

Some felt this acutely. William Blake loathed Newton and urged a renewal of "fallen, fallen light." Other Romantics celebrated light in symphonies, paintings, and poems, yet scientists continued to probe. Bouncing beams through dark chambers, measuring first by "candle power," later with lumens and watts and joules, they dimmed divine radiance and opened a new debate about light. Particle or wave? But even after Einstein muddled the question, even as today's scientists craft miracles in optics labs, light still performs. Tens of thousands gather for the summer solstice at Stonehenge. Festivals of light illuminate Berlin, Chicago, Hong Kong, Ghent, Amsterdam, Lyons . . . The anatomy of light—particle *and* wave—is a staple of science classrooms, yet no equation has dimmed sun or moon, and no prism rivals the rainbow.

Light attempts to reconcile the battles between science and humanities, between religion and doubt, between mathematics and metaphor. Like sunlight, which as Thoreau noted "is reflected from the windows of the almshouse as brightly as from the rich man's abode," this radiant history illuminates all those in love with light. It focuses equally on the genius of Newton and Dante, the eloquence of equations and scripture, the faith of the Qur'an, the Upanishads, and the Bible. Of light's devout disciples, I ask not "who" studied light but "why." Why were these believers enamored of light and what did their faith add to human consciousness? Of light's students I ask not "what" but "how." How did scientists determine the nature and tame the power of light? And of those who made light their muse, I ask only that they be read and seen as if for the first time.

Yet the prime mover of this history is neither experiment nor eloquence but awe. The story begins at the approach of dawn. The long night is ending. Daybreak is near. A glimmer touches the eastern horizon. Hail, holy light—particle, wave, and wonder.

Part One

Where is the way where light dwelleth,
and as for darkness, where is the place thereof?

—JOB 38:19

CHAPTER I

"This Light Is Come": Myths of Creation and First Light

May the warp be the white light of morning,
May the weft be the red light of evening,
May the fringes be the falling rain,
May the border be the standing rainbow.
Thus weave for us a garment of brightness.

—THE TEWA PEOPLE,
"SONG OF THE SKY LOOM"

Regardless of any weather, heedless of any clock, summer light comes early to England's Salisbury Plain. Well before sunrise, the enormous, vaulted sky flushes gray then white, salmon then yellow. Light floods the rolling landscape seventy miles west of London, spreading across a green as broad and smooth as an Alpine meadow. Here on this expansive plain the ancients came to chart light, to pay homage, to feel its power. Here light's long story is written in stone.

On most nights, cars and trucks plow the four-lane A303, their drivers paying scant attention to roadside shadows. The hulking granite slabs for which the Salisbury Plain is world famous stand alone in the dark. Stars whirl above and the moon draws little notice. But this night is different. Though the sky is still black and brooding, some twenty thousand people have gathered at Stonehenge. Swarming around the circle of stones, dancing in the glow of distant floodlights, the crowd includes an array of colorful characters. Some are pagans, dressed as if the term "hippie" did not suggest a period piece. In painted faces and flowing robes, they will drum and chant all night. *All* night. Others are neo-Druids—men with bushy beards and white tunics, women with leafy crowns and flowered dresses—holding forth with neo-ancient wisdom. And thousands more are just college students or tourists with good timing, who heard that on this particular morning, in this particular place, dawn is tantamount to magic.

In defiance of our floodlit age, the annual summer solstice at Stonehenge has become a Woodstock of Light. Each June 20, shortly after sunset, crowds begin streaming in from the "car park" to assemble around these stones. As twilight fades and night descends, music drifts through the semidarkness, reggae mixing with still more drumming. Come midnight, revelers blow bubbles or stand with arms outstretched in ecstasy. On the fringes of the crowd, carnival-colored hoops spin. Light blinks from wands, and glitters on headgear ranging from top hats to sombreros. Human joy seems palpable here. Light is coming. Forget war and poverty, hunger and despair. Light is due.

Watches and phones now read three thirty A.M. Overhead, the sky is sprinkled with stars. The Big Dipper wheels; the night rolls on. Encircled by the massive slabs, crowds swirl to the echo of drums. Small groups chant and cheer; lone dancers twirl and glide. Huddling for warmth or pacing away the hours, those less in light's thrall begin to wonder whether dawn will ever come. Then, toward four A.M., the eastern horizon softens to a royal blue. Clouds appear, stacked like shelves—gray over white over pink. A palpable surge passes through the crowd. Above the stones, the sky slowly pales. One by one the stars go out. All eyes focus east, fixing on a brightening smear of yellow. Hands rise in greeting. Primal shouts soar. The horizon seems driven toward this moment, as if, like the E. E. Cummings poem, "this is the sun's birthday." Yellow now silhouettes the black skyline of earth, framing the heads turned toward it, waiting, waiting. Finally at four fifty-two A.M., a spear of light bursts over the earth's edge. The sun gilds the stones, sweeping down their surface, catching each lifted face full on. Raucous cheers arise. Light! Light again! As it has for more than four thousand years, light has returned on schedule, on the summer solstice, to Stonehenge.

Just as no one is certain who built Stonehenge, the mysteries of light's creation remain. How was First Light made? What deity or force of nature flipped the switch? Was there fanfare, the universe itself bursting with pride? In the last century, science has discouraged such romanticism, its Big Bang replacing primeval wonder with protons and quarks. Yet when the builders of Stonehenge watched light return, they saw little need to explain its origin. Like all peoples, they already had a creation story. And as in all creation stories, light was its preface.

Because all good came from it and most evil came from its absence, primitive people did not study light—they worshipped it. Envisioning light's birth, our distant ancestors called on myth and miracle. Night was their muse, night, when whole tribes clustered around fires, weaving creation stories more fabulous than the diamonds above. Giants prowled the earth in some stories. In others, a prime mover sat alone in a universe "without form, and void." Darkness was the default mode—"darkness, blinding darkness," "darkness swathed in darkness," "darkness upon the face of the deep." When time had not yet begun, anything seemed possible. Dogs spoke. Women fell from the sky. Children were split in half, and First Light came from God's eyes, teeth, armpit, even His vomit.

The myths of origin are as vital as any stories ever told. Myth, said the celebrated mythologist Joseph Campbell, is "society's dream." And creation myths, wrote Marie-Louise von Franz, a disciple of Carl Jung, are "the deepest and most important of all myths," because they reveal "the origins of man's conscious awareness of the world."

Whether a creation story starts with "In the beginning" or Coyote and Sun-Woman, everyone knows one. Every child asks for one. Many cultures have more than one, making the number of creation myths exceed the number of peoples on earth. Despite this abundance, however, mythologists categorize creation myths into five types: 1) Earth Diver, in which gods dove beneath dark waters to dredge up the first land; 2) World Parent, in which gods, male and female, gave birth to all living things; 3) Ex Nihilo, wherein a god formed the universe "out of nothing" (the meaning of the Latin phrase); 4) Emergence, charting the first humans' ascent from a lower world; and 5) Cosmic Egg, in which a primordial egg hatched to give birth to a god, who took things from there.

In all five archetypes, light holds a privileged place. Creation myths make the birth of flora and fauna a messy affair involving mud and slime, murder and incest, sin and salvation. But the creation of light is universally regarded as a gift. In Zuni Indian myth, First Peoples emerge from an underworld into blazing light. An Orphic Greek hymn describes an "all-spreading splendor, pure and holy light." In the Finnish creation legend, *The Kalevala*, an egg opens and

> the fragments all grew lovely.
> From the cracked egg's lower fragment,

Rose the lofty arch of heaven
From the yolk, the upper portion,
Now became the sun's bright lustre;
From the white, the upper portion,
Rose the moon that shines so brightly;
Whatso in the egg was mottled,
Now became the stars in heaven . . .

The god or gods who made the first people often had misgivings, leading to mythology's litany of floods and false starts, yet no primal myth describes a Second Light. Eons before it streamed into Gothic cathedrals or hummed ruby red out of the first laser, light was perfect. In the Hindu Upanishads, each primordial sunrise brought "shouts and hurrahs." The Kono people of West Africa speak of light dawning when a bird sang. And then there is Genesis 1:4: "God saw the light, that it was good." Flawless First Light let creation unfold, much as light still does in the world each dawn creates anew.

Nearly three hours before the sun rises over Stonehenge, dawn comes to the Olduvai Gorge in Tanzania. Unlike England's Salisbury Plain, Olduvai, home of our oldest human ancestors, is rimmed by rugged buttes that block the sun long after night begins to fade. Throughout this parched gorge and across the surrounding valley, the first people pondered the origin of light. The earliest creation myths here have largely been supplanted by one told in Swahili: "Before the beginning of time there was God. He was never born nor will He ever die. If He wishes a thing, He merely says to it, 'Be!' and it exists. So God said: 'There be light!' And there was light . . ."

Like East Africa itself, the story is heavily influenced by the Qur'an, which speaks of a six-day creation, including Adam and Eve and a garden. To find older creation stories, follow the light westward into what Europeans once feared as a "heart of darkness." From Olduvai, it takes dawn just an hour to spread across deep blue Lake Victoria and filter into the lush jungles of the Congo region. Here Bushongo tribesmen tell of the god Bumba. Creation began on the day when Bumba, alone on dark waters, writhed in pain. His stomach seemed to be splitting. Retching, flailing, Bumba felt his entrails begin to empty. Finally, with a great groan and flash, Bumba "vomited up the sun." As sunlight spread, it dried the waters. Ridges of land appeared. Climbing onto dry earth, Bumba bent to the ground and regurgitated moon and stars. Beneath their radiance, the queasy god then vomited the leopard, the crocodile, the tortoise, the fish, and finally, man.

This Bushongo myth is one of many "out of nothing" stories in which light comes from the body of a god. First Man and First Woman might have been fashioned from mud or wood, but light's perfection suggested that it was a piece of the godhead. If vomit seems too earthy, consider a tooth. On the Gilbert Islands in the South Pacific, the Maina people speak of the god Na Arean who sat alone in the dark atop "a cloud that floats in nothingness." One day, as Na Arean brooded, a small man burst from his forehead. "You are my thought," Na Arean said. This god and his thought remained in darkness until the little man stumbled and fell. Then Na Arean yanked a hollow tooth from his jaw, thrust it into the land he had made from mud, and behold—light streamed through. Elsewhere, creation myths find light bursting from beneath God's armpit (Kalahari Bushmen) or lighting the universe when he laughed (Northern Egypt). Yet the most common source of primal light was God's eyes.

A Chinese story speaks of the Great Creator, Phan Ku. Hairy, with horned tusks, Phan Ku was a giant who grew ten feet a day, lived for eighteen thousand years, and made the universe from of his hideous body. Phan Ku's tears became the Yangtze River, his breath begat the wind, his bones became rocks, his voice, thunder. And when the Chinese looked into the sky, they saw Phan Ku's left eye shining by day, his right eye beaming by night. Early Egyptians saw a similar source. Up and down the Nile, stories of human origin vary, but most agree that the sun god Ra created light. "I am the one who openeth his eyes and there is light," Ra proclaims. So long as Ra's eyes are open, daylight prevails. "When his eyes close, darkness falleth." But as in many cultures where light meant life, Egyptians saw demons lurking beneath the horizon. The snake (or sometimes dragon) Apep was Ra's mortal enemy, seizing sunlight each evening and battling on through the night. Dawn over the Nile meant not just light and warmth but victory. Ra had prevailed again.

Spun from a god, light was the seed of creation. Some cultures, however, saw light as too ethereal to have an origin. The tortoise, the snake, First Man and First Woman were made in the beginning, yet many creation myths unfold beneath a light that, like some gilded frame for the universe, simply *is* and has always been. Given light's omnipresence, it was darkness that demanded explanation.

As it does at Stonehenge, dawn dominates the Banks Islands, a small archipelago in the South Pacific. Above white sand . . . beaches, beneath

With the sun streaming from his eyes, the Egyptian sun god Ra was one of many mythological deities explaining the origin of First Light.
iStockphoto.com

fluffy clouds, the sky is rimmed by ocean. At thirteen degrees of latitude south of the equator, the islands are washed by a light whose daylight hours vary little throughout the year. Circling any island, you can watch light dawn over water each morning and disappear over water at dusk. Perhaps light's ubiquity explains why Banks Islanders talk of a time when night never fell.

Banks Islands creation myths center around the god Qat. Born from a bursting stone—some say the stone still stands in a certain village—Qat set about creating the world. He carved people out of wood, then made jungles and beaches, volcanoes and coral reefs, pigs, fire, and rain. The only thing Qat could not make was darkness. "There's nothing but light all the time,"

Qat's brothers complained. "Can't you do something?" But light remained relentless until Qat heard a new word. It seemed that in the neighboring Torres Islands something called "night" descended. To please his brothers, Qat set out in his canoe to get some of this "night." Paddling across the ocean, he reached the Torres Islands, where he traded a pig for a piece of night and a few roosters. Then the god headed home. Within hours, the sky above the Banks Islands began to dim. Qat's brothers were terrified.

"What is spreading and covering the sky?"

"This is night," Qat said. "Lie down and keep quiet."

Qat's brothers lay down and soon felt sleepy.

"Are we dying?"

"This is sleep."

Night might have lasted forever but for the roosters. When they began to crow, Qat grabbed a sharp rock and slit open the sky. Light returned to the Banks Islands, as it has every morning since.

Dawn moves west from these islands, speeding over the waters, arriving an hour later at the Australian Outback. Here on this red-rocked desert, Aborigines speak of Dreamtime, a state of being that stretches from creation to infinity. When Dreamtime began, elders say, the sun never set. Scorched and nearly blinded, the people of Dreamtime sought relief. Finally, the god Norralie cast a spell: "Sun, sun, burn your wood, burn your internal substance and go down." And ever since, the sun has vanished each night, letting the Outback cool.

Unbroken light also fills the myths of the Incas, the Mayans, and several Native American nations. Yet the light that has always been has not always been in its proper place. For the Miwok Indians of central California, first peoples lived in gloom, the only light coming from a glow in the east, said to be the home of Sun-Woman. Longing for their own light, the Miwoks sent Coyote-Man to fetch Sun-Woman. He found her sitting, radiant in a garment of abalone shells. Sun-Woman refused to budge, however, so several Miwoks came, tied her up, and brought her back to share her light. In similar stories, various Eskimo peoples speak of a light that only emerged when a bird pecked a hole in the dark dome, or when the trickster Raven put the sun in a bag and delivered it. Arizona's Yuma nation had an easier time finding light. Their creator spit on his thumb, polished a portion of the black sky, and the light shone through.

Stories of First Light are told and retold in oral traditions that confined each story within its tribe, each First Light to its farthest horizon. Yet once writing emerged in Mesopotamia, then spread to China and India, the world's major religions committed their myths to parchment and tablet. Their radiant eulogies would be as enduring as any other human creation.

No celestial light dominates the earth like the dawn over India's Ganges River plain. Night after night in the holy city of Varanasi, candles float across the inky river, giving darkness a soft amber glow. Oil lamps flicker beneath the wall of arches and turrets lining the Ganges. As sunrise approaches, a fan of radiance spreads above this ancient city, also known as Kashi, "City of Light." Now as hundreds bathe and bow to the eastern horizon, yellow spreads into orange, orange into red. Then, suddenly, a shimmering, pulsing, bursting sun spills over the horizon, daubs its reflection on the waters, and rises to scorch another day.

In the complex web of Hindu scriptures, whose Vedas are the world's oldest holy texts, a myriad of creation stories overlap. All agree that creation began in darkness, or, as the Rig Veda says, "darkness swathed in darkness." Piercing this black void, First Light had many sources. Some say the mighty Prajapati, later known as Brahma, begat light and dark. He breathed, sending Devas ("the shining ones") to light the sky. Other stories describe the primal man, Purusa, sacrificed by the gods, whose body became the universe, his mind the moon, his eye the sun. Hindu scriptures incorporate the full range of creation archetypes, from Earth Diver to Ex Nihilo, from Cosmic Egg to World Parent. In each, however, light brings pure joy. In one of its many hymns to the dawn, the Rig Veda proclaims, "This light is come, amid all lights the fairest; born is the brilliant, far-extending brightness . . ."

The same Ganges sun that floods Hindu shrines also lights Buddhist temples, whose sand-castle stupas silhouette the eastern horizon. Feeling the sun's warmth inside them, the Buddha and his followers on this plain saw light born both in the sky and in the soul. Buddhist scriptures say little of creation and nothing about a creator. In lieu of beginnings, the Buddha spoke of a universe that perpetually expands and contracts, creating a wheel of life, a samsara, that cycles birth, death, and rebirth. Asked who made this wheel, the Buddha answered, "Inconceivable, O Monks, is this Samsara, not to be discovered in any first beginnings of beings." Yet on the origin of light Buddhism is more certain—it once came from within.

"There comes a time," the Buddha told his followers, "when sooner or later after a long period, this world contracts. At a time of contraction, beings are mostly born in the Abhassara Brahma world. And there they dwell, mind-made, feeding on delight, self-luminous, moving through the air, glorious—and they stay like that for a very long time." Self-luminous. Glowing with inner light. With no sun, moon, or stars, these beings floated through a realm lit only by themselves, what the Buddha called a "World of Radiance." After another "very long period of time," the glowing beings were surrounded by "savory earth [that] spread itself over the waters." Tasting the fresh earth, finding it as sweet as honey, they feasted on whole handfuls. "The result of this was that their self-luminance disappeared . . . moon and sun appeared, night and day were distinguished, months and fortnights appeared, and the year and its seasons." Born within, light fled, leaving us drawn to sun and moon, the light we once contained. Two thousand years before Milton, the Buddha's World of Radiance was Paradise Lost.

The world's scriptures contain a wealth of stories about First Light, but the god who created the world in six days has only one. The Book of Genesis dates to about 600 BCE. Its opening lines have been recited in catacombs and cathedrals, alone in quiet corners, and by astronauts orbiting the moon. Still they bear repeating.

In the beginning, God created the heaven and the earth,
And the earth was without form, and void; and darkness was
upon the face of the deep.
And the Spirit of God moved upon the face of the waters.
And God said, "Let there be light." And there was light.
And God saw that light, that it was good. And God divided the
light from the darkness . . .

From biblical days until the age of Darwin, Genesis defined First Light in Western culture. Its signature phrase, "Let there be light," has inspired artists and composers, poets and authors, scholars and saints. The Latin translation, "Fiat lux," is the motto of dozens of colleges and universities. Contemporary scholars argue endlessly about Genesis, its origins, its morals, its meaning. Translators vary on much of the text, yet from the King James to the Revised Standard to the New King James, all agree on this single phrase. Every translation reads: "Let there be light."

Created out of nothing, First Light in Genesis is as mysterious as it is divine. It may seem like another light among many, yet it bears one vital distinction. Every other primordial light comes from sun, moon, or stars, or else from gods embodying these. Yet here, at the end of the beginning, we find light made from neither sun nor moon. God begets it with His first words, then moves on to create heaven and earth. Only on creation's fourth day does He make "the two great lights; the greater light to rule the day and the lesser light to rule the night." Some consider this one of those "gotcha" moments atheists love to throw in the face of true believers. But it may be more.

Just as God separates light from dark, Genesis separates light from its celestial sources, making light a force of nature. Even when "the greater light" is behind the clouds or "the lesser light" is on the far side of the earth, this entity called light fills creation. Light, Genesis suggests, is more than the sum of sun, moon, and stars. While stopping short of considering it energy—that would take centuries of science to reveal—Genesis recognizes light as the essence of the universe. God creates it, calls it good, separates it, and only three "days" later (gotcha!) does He see fit to create sun and moon. A more practical god might have made these orbs first in order to, as Genesis notes, "let them be for signs, and for seasons and for days and for years." But this god's light is a tool, the first tool of creation.

Linking myth to behavior is a risky business, a chicken-and-egg question. Do cultures behave according to their myths, or do they cling to myths that explain their behavior? The debate will never end. But by embracing a creation story that distinguishes light from sun and moon, perhaps the Western world (including Islam) freed itself to study light as an entity, not just an inspiration. The anthropologist Bronislaw Malinowski wrote, "An intimate connection exists between the word, the mythos, the sacred tales of a tribe on the one hand and their ritual acts, their moral deeds, their social organization, and even their practical activities on the other." For Malinowski and many other mythologists, myth is "not merely a story told, but a reality lived." Though the connection is not definitive, the light created on the first day of Genesis will, as its story unfolds, be most fully examined in cultures whose god did not emit radiance or slice open the sky, but said simply, "Let there be light."

Sweeping around the globe, light burst forth. The wealth of primal myths suggests that light came to human consciousness from both within and without. Some cultures were content to have it spring from a god's body. Others regarded light as euphoric, its creation as joyous as laughter or song. But all First Light, advancing across continents, oceans, and deserts, was greeted with reverence, awe, and no small measure of mystery. "Was there a below?" the Rig Veda asks:

Was there an above?
Casters of seed there were, and powers;
Beneath was energy, above was impulse.
Who knows truly? Who can here declare it?
Whence it was born, whence is this emanation?

And the answers, wholly mythological yet wholly believed, began humanity's long journey out of darkness.

CHAPTER 2

"The Thing You Call Light": Early Philosophers from Greece to China

"You surely realize that even in the presence of color
Sight will see nothing, and the colors will remain unseen,
Unless one further thing joins them . . ."
"What thing do you mean?"
"The thing you call light."

—PLATO, *THE REPUBLIC*

Like those who later stood on their shoulders, the first students of light were a curious bunch. One insisted the universe was made of water and that magnets had souls. Another claimed to be a god, proving it by hurling himself into the fires of Mount Etna. Only a sandal was found, or so legend said. Some spoke of a chariot passing through the Gates of Night and Day, driving on through pure light. Several beamed sunlight off curved mirrors, focusing its rays on small sticks, cheering when they burst into flame. Many refused to eat beans.

But whereas others saw light as holy, these early students insisted that it could be studied. Light was not a piece of god, they said. Then "whence is this emanation?" as the Rig Veda asked. Light came from the eye, some argued. No, it came from the object seen. The eye. The object. The eye . . . The argument seemed trivial to some, yet its secular tone enabled all the discoveries of light in the two dozen centuries since.

Aristotle called these philosophers *phusikoi*, or "students of nature." From the Greek, we get the term "physics" under whose umbrella the study of light still proceeds. The quest of the *phusikoi*, Aristotle wrote, was to know "the cause of visible things," and "what things are made of." These were the days—four, five, six centuries before the birth of Christ—when little besides scripture had been written down and all things demanded explanation. Scattered across islands washed by the Mediterranean and burnt by the sun, the first Western philosophers—like their counterparts in India and China—set out to structure the known world. Some started by explaining the soul, others by asking how we should live, whether to trust our senses, and what, other than the whims of the gods, explained natural phenomena. Among the first phenomenon they sought to explain was light.

Surely it was made of rays. Or perhaps atoms. It must create a sheen that rippled some invisible substance between eye and object. Question led to question, but this much seemed certain: Unlike sound, whose measured pace was plain to anyone shouting across a field, light did not "travel." Sound paused between collision and crunch, but there was no such delay with light, not even between opening one's eyes and seeing the stars. Light had to be instantaneous. And even those who did not brood on "what things are made of" found it fascinating. Painters wondered about color and shadow. Healers sliced open animals' eyes and marveled at the miracle of vision. Astronomers charted "the pure torch of the shining sun" and "the wandering works of the round-eyed moon." And slowly, study by study, the divine light of creation met its match in human curiosity.

Light seemed instantaneous, yet the first theories about it traveled no faster than a sailing ship. Between the Greeks' earliest conjectures about light and their last textbook on optics stretched seven hundred years, the same as the span that separated Dante's divine light from quantum theory. During these seven centuries, the sun rose and set a quarter million times, making light one of nature's dualities—hot and cold, male and female, light and dark. Most who thought at all about light fell back on stories of creation. In his *Theogony*, the Greek poet Hesiod told of the god Chaos, born in darkness, who mated with Eros to spawn Erebus, the god of night, and Aether, the god of light or pure air. (More about Aether later. Much more.) The blind poet Homer reveled in describing light. *The Iliad* and *The Odyssey* abound with "luminous epiphanies of the gods," including Zeus's lightning and Athena's descent from Olympus "in a trail of light." Homer's heroes don glowing helmets or appear beneath glittering clouds. Homeric radiance, with its "rosy-fingered dawn," defined light throughout Greece until about 600 BCE. And then, emerging ex nihilo like the creation in Genesis, a few of the curious began talking of light as something other than divine.

Along the coast of modern-day Turkey, the first Greek philosophers, the pre-Socratics, examined light by looking to the skies. In 585 BCE, Thales, so fascinated by light that he once fell into a well while gazing skyward, predicted a solar eclipse. Modern scholars doubt this legend, yet Thales recognized that moonlight was reflected sunlight, and that, while giving the gods their due, light could be explained without them. One plodding century later, Anaxagoras took up the study. Asked for what purpose he had been born, Anaxagoras responded, "For the contemplation of the sun, and moon, and heaven." The sun, he said, was a mass of burning iron from which pieces might fall. For this proclamation Anaxagoras was tried, convicted, and banished. Perhaps all things demanded explanation, but there were limits. Another few decades passed before light began to fascinate a student as obsessed with himself as with "the cause of visible things."

On the southern shore of Sicily, high above Homer's "wine-dark sea," stand the western fringes of ancient Greece. Perched on a hill behind a treasure trove of ruins is the Italian city of Agrigento, with its requisite churches, piazzas, and red-tile roofs. Tourists who have not done their homework are sometimes surprised to find Greek ruins in Sicily, yet Agrigento's Valle dei Templi (where I slept alone one Christmas Eve, gazing at Orion above the pillars) plays host to rich remnants of Greek glory. Dotted with palms and olive trees, this sepia-toned city stretches for blocks. Scaled-down Parthenon look-alikes—the temples of Juno, Heracles, Zeus, and Vulcan—rise above adjacent fields. Crumbled roofs leave shells of buildings open to the sky. Steps are broken, paved stones uneven, whole blocks laid waste by time. It was here, in what he called "the great town of the tawny Acagras," that the philosopher Empedocles first posed the question: What is this thing called light?

Though mere fragments of Empedocles' writings remain, legends about him suggest the eccentricity of light's earliest students. Striding these streets in a purple robe and golden girdle, he let his hair grow long, told admirers of his former incarnations as a bush, a bird, a fish, and announced that in this life he was a seer or perhaps a god. Stories about Empedocles spread throughout the Mediterranean. He had raised a woman from the dead, calmed a murderer by playing a lyre, blocked gale-force winds by stretching animal skins across a mountain pass, and thrown himself into fire and emerged unscathed. Scholars now doubt that he leapt into Mount Etna, yet it seems entirely in character. These were the quirks of light's first sage, quirks hiding a seasoned view of nature and man.

"Alas, poor race of mortals," Empedocles wrote, "unhappy ones, from what conflicts and what groans you were born . . . I too am now one of these, an exile from the gods and a wanderer, having put my trust in raving strife." Strife, Empedocles believed, formed another of nature's dualities, its opposite being Love. The two waged a tug-of-war for control of the earth, a Greek yin-yang whose believers disappeared shortly after Empedocles died in 435 BCE. More enduring was his list of nature's four prime elements: earth, air, fire, and water. Light, Empedocles said, was fire, one that came from the eye.

> As when someone, intending a journey, prepares a light,
> A flame of flashing fire through the winter night,
> Fitting a lantern as protection against all the winds,
> Which stops the breeze when the winds blow,
> But the light passes through to the outside, inasmuch as it is finer-textured,
> And illuminates the ground with its tireless rays.

Some scholars see more metaphor than science in this fragment, yet others insist that Empedocles defined light as optical fire "imprisoned in the membranes" and protected by aqueous fluids—"the deep water which flows around." Like light from a lantern, the eye projected its flame *toward* objects. Later, however, Plato quotes Empedocles as claiming that light pours *into* the eye. So did light come from eye or object? The debate, initiated by Empedocles arguing with himself, had begun. By 410 BCE, an alternate theory had emerged, suggested by a common visible phenomenon.

Flecks of dust float in a sunbeam, drifting, dancing. What do they say about matter, about the light in which they swim? Perhaps everything on earth, from an ordinary table to the tallest mountain, is composed of tiny pieces, indivisible, called atoms. The philosopher who promulgated this theory, Leucippus, soon applied it to light. All objects, Leucippus wrote in the fifth century BCE, emanate wafer-thin particles of light. Shaped like the objects that emit them, these particles create simulacra that travel *to* the eye. The philosopher Epicurus explained this in a letter. (Yes, these men wrote letters about light.) "There are images or patterns that have the same shape as the solid bodies we see but are very thin in texture . . ." Epicurus wrote. "We call them eidola." Shimmering and spectral, eidola (singular: eidolon) proved that light could not possibly be a fire in the eye. "For external things," Epicurus continued, "would not stamp on us their own nature of color and form through the medium of the air which is between them and us, or by means of rays of light . . . so well as by the entrance into our eyes or minds." By 400 BCE, the theory of eidola reached Athens.

Throughout Greece's Golden Age (480–323 BCE), Athenians adored light. Each morning they thanked Eos, the goddess of dawn, and sometimes Phosphoros, the "light bearer," whom the Romans knew as Lucifer and we call Venus. Each evening, Athenian homes glowed with lamps burning olive oil. Elaborate torch-lit ceremonies accompanied births, marriages, and deaths, while every four years, a torch that was trotted through the streets heralded the Olympic Games. Even Greeks steeped in myth and ritual could ponder light's mercurial behavior. A spoon in a glass seems to bend at the water line. Thick glass makes objects look bigger. The rainbow comes and goes. Who could explain these marvels if not those curious characters with nothing better to do than wonder about light?

Plato did not care much about light. As the founder of his own academy, teacher of Aristotle, and interpreter of all that Socrates had said, he could only touch on the subject. Eye vs. Object? In his dialogue "Timaeus," Plato tried to have it both ways. The gods, says the sage Timaeus, put a special fire into the human eye, a fire that would not burn. "And the pure fire which is within us and related thereto they made to flow through the eyes in a stream smooth and dense." By day, Timaeus posits, light from our eyes meets the light reflected by all objects. At night, however, fire from the eye finds no reflected light. Unable to see, we sleep, our dreams lit by that special fire trapped behind closed eyelids.

In Plato's *Republic*, Socrates treats light solely as metaphor. In lieu of the *phusikois'* theories, he uses "the thing you call light" to evoke knowledge, honor, and goodness. For Socrates, the difference between knowledge and ignorance is the difference between seeing by day and groping by night. "You can look at the soul in the same way," he continues. "When it focuses where truth and that which is shine forth, then it understands and knows what it sees, and does appear to possess intelligence. But when it focuses on what is mingled with darkness . . ." Expanding the idea, Socrates gave light its most enduring allegory. Imagine a cave where men chained in darkness see only shadows on the wall. When one slave is freed, he emerges into the sunlit world. Blinded by its light, he assumes this outside realm to be unreal. Only with time does he come to understand. Returning to the cave, he tells the others, but none believe him. Yet the man will never be the same, having found "the lord of light in this visible world, and the immediate source of reason and truth in the intellectual." The famous Allegory of the Cave lit the candle that would forever link light to knowledge, a candle soon handed to Plato's most famous pupil.

Aristotle scoffed at all previous explanations of light. The most quarrelsome of the *phusikoi*, he posed a question that should have decided Eye vs.

Object. If the eye emits light, "Why should the eye not be able to see things in the dark? . . . It is, to state the matter generally, an irrational notion that the eye should see in virtue something issuing from it." Enough metaphors and guesswork, Aristotle said. Light was neither fire, nor atoms, nor shimmering eidola. Light was an "activity—the activity of what is transparent," clearing a path through the air to let images pass. "It is better, instead of saying that the sight issues from the eye and is reflected, to say that the air, so long as it remains one, is affected by the shape and color." Having defined light, Aristotle explored its showiness—the flash of a stick hitting water, haloes around the moon, and why we "see stars" when smacked in the head. Question led to question. If light came from objects, what conveyed each ray to the eye?

The *phusikoi* knew that sound rippled the air, hence *something* between eye and object must be rippled by light. "Vision occurs when the sensitive faculty is acted upon . . ." Aristotle reasoned. "There only remains the medium to act on it, so that some medium must exist; in fact, if the intervening space were void, not merely would accurate vision be impossible, but nothing would be seen at all." Choosing a name with deep roots in ancient Greece, Aristotle called this medium "aether."

Purple-robed Empedocles had defined four elements—earth, air, fire, and water—but Aristotle added a fifth, aether. The name came from Aether, the Greek god of—depending on the source—Light, Brightness, or Pure Air. Because nature abhorred a vacuum, Aristotle insisted, a transparent, mellifluous, and perfect aether must fill the heavens and all space between the stars. Once Aristotle defined it, aether also filled the vast time and space between ancient Athens and a basement outside Cleveland, Ohio, where, in 1887, the existence of aether would finally be disproven. The experiment that ended speculation about aether would pave the way for light's conquest of the twentieth century. But in the millennia between, ether, stripped of its *a*, would be considered as essential to light as light was to life itself.

Whether divine or ether-borne, light is the most democratic of all natural phenomena. Unlike land and water—scarce in some places, abundant elsewhere—light is uniformly bestowed throughout the earth. Like a beach ball revolving in a flashlight beam, the surface of the planet is bathed equally by light. Due to cloud cover, the sun may not shine as often in Moscow as it does in Mexico, and thanks to the earth's tilt, days shorten and lengthen with the seasons. Yet in a given year, the total hours of daylight are the same everywhere and have been since light was first fetched by Coyote-Man, vomited by Bumba, or begat by God. It is not surprising, then, to find parallel discoveries about light around the globe. What seem strange are the contrasts.

Flecks of dust float in a sunbeam, drifting, dancing. And two hundred years before the Greeks saw the world as made of atoms, dust inspired the same theory in India. In the sixth century BCE, an anonymous observer from India's earliest scientific school, the Vaisheshika, wrote: "The mote, which is seen in a sunbeam, is the smallest perceptible quantity. Being a substance and an effect, it must be composed of what is less than itself . . . This again must be composed of what is smaller; and that smaller thing is an atom." The Vaisheshika sutra did not ask whether light came from eye or object, instead defining light by its properties: "Light is colored and illumines other substances, and to the feel is hot: which is its distinguishing quality." Regarding light and heat as "one substance," the sutra posited two kinds of light, "latent or manifest." One was seen, the other felt. "Fire is both seen and felt. The heat of hot water is felt but not seen; moon shine is seen but not felt. The visual ray is neither seen nor felt." Terrestrial light was earthy, celestial light, "watery." When combined, they made gold, whose "chief ingredient is light . . . rendered solid by mixture with some particles of earth."

India provided parallels to Western thought; China offered contrasts. While pre-Socratic Greeks argued over light, the Chinese studied shadows.

- "Shadows are two . . . when two lights flank one light, whatever is illuminated by one light is in shadow."
- "When the post slants, the shadow is shorter and bigger; when the post is upright the shadow is longer and smaller."
- "[When light passes through a hole in a wall,] the turning over of the shadow is because the criss-cross has a point from which it is prolonged with the shadow . . . the light enters and shines like the shooting of an arrow."

These observations come from the Mo Jing, a study of space, time, geometry, logic, "universal love," and light. From the fourth to the second century BCE, the practical Mo Jing complemented the esoteric wisdom of Confucianism and Taoism. Whereas Confucius, like Socrates, preferred metaphor—"It is better to light one candle than to curse the darkness"—the Mohists experimented with candles, mirrors, and the first pinhole camera. Holding up assorted objects, they noticed how shadows behaved in multiple light sources and mirrors before mirrors. Planting sticks in the ground, they studied lengthening and shortening shadows, delighting in how light played with darkness. Mohists knew that light traveled "like the shooting of an arrow," but they cared little whether the arrow was shot by eye or object. Instead, using the holistic approach that would continue to characterize Chinese thought, they pondered how light affected the whole human body. They studied light's impact on the *zhi* ("understanding" or "intelligence") and the *xin*, the organs governing sensory perceptions.

By 220 BCE the Chinese knew everything the Greeks knew about light, and perhaps more. Then the Qin Dynasty emerged, ending two centuries of political upheaval. The Mo Jing was one of many texts that were burned. In a culture more concerned with "inner light" than the beams that came from eye or object, few stirred the embers. Concerned with dualities and balance, Taoists and Confucianists saw light entwined with darkness, not a "thing" to be studied on its own. The Taoist philosopher Zhuangzi cast aspersions on such study: "If a thing has no form then numbers cannot express its proportions, and if it cannot be encompassed, then numbers cannot reach to infinity to express it." Like infinite knowledge, Zhuangzi concluded, light and shadow could never be mastered. By the second century CE, some in China spoke of "bright window dust," the same floating specks noticed in Greece and India, yet the dust was compared to gold, to vapors, and to *chi*, the body's energy. Natural light continued to illuminate China as democratically as it fell upon the rest of the globe, yet under succeeding dynasties it spawned metaphors rather than measurement. "That which lets now the dark, now the light appear is the *Tao*," says Taoism's *Book of Changes*, the *I Ching*. While the Chinese looked inward, the Greeks were building the brightest light in the world.

In 331 BCE, Alexander the Great began mapping out a city worthy of his name. Across a broad plain fronting Egypt's Mediterranean coast, Alexander's assistants laid out streets in lines of barley flour. On one edge of the new city would stand the Gate of the Sun. At the other would rise the Gate of the Moon. And a mile off its shore, on a small islet called the Pharos, would be a lighthouse. Any ordinary light atop a tower would have warned ships of dangerous shoals, yet in stone after stone the lighthouse of Alexandria rose. To one hundred feet. To three hundred. To four hundred fifty feet, as imposing as a forty-story building and, after Egypt's Great Pyramid, the world's tallest structure. Arches were added, and balconies and bronze carvings. Huge mirrors atop the tower reflected a blazing furnace, sending out a beam visible for miles.

The light of the Pharos reached far beyond its rays, becoming the symbol of the greatest center of learning the world had yet seen. With a population of a half million—twice that of ancient Athens—Alexandria contained more light than any other city on earth. Between its gates stood civilization's first museum, dozens of palaces and temples, and a library said to hold all existing knowledge. The light of Alexandria drew the best thinkers of the pre-Christian era. Here Eratosthenes used noontime shadows to calculate the circumference of the earth—accurate to within 2 percent. Here, long before Copernicus, Aristarchus described a sun-centered universe. The first steam engine was made here, as were the first medical dissections and the first translations of the Old Testament. And here beneath the beam of the Pharos, the Greeks took light where the Chinese dared not—into the realm of numbers.

Little is known about Euclid, yet the name still resounds in classrooms worldwide. Euclid's *Elements*, devised in Alexandria, remains the basis of modern geometry. In about 300 BCE, Euclid applied his geometric theorems to light, founding the science of optics. All subsequent study of light dates to Euclid's simple opening statement: "The ray issues from the eye in straight lines." *From* the eye. But Euclid did not delve further into sources. Forget eidola, particles, even ether. Euclid's *Optica* treats light as angles and rays and points, as much "a thing" as an arrow. In fifty-eight propositions, each more complex than the last, the father of geometry geometrized light. Rays of light form cones, Euclid wrote. The size of each determines spatial and depth perception. Objects not falling within vision's cone are not seen. Euclid backed up each statement with a proof that concludes, "And that is what we wished to demonstrate." Though not recommended as bedtime reading, Euclid's *Optica* sketched the myriad ways in which light ricochets in patterns that resemble Japanese origami, unfolded.

Euclid's *Optica* laid the groundwork for the brilliant lights of our age. The backlighting of a smartphone, the laser that etches silicon chips, all the dazzling light shows—none would have been possible without Euclid. The great geometer also examined mirrors. The Greek word for "mirror" is *katoptron*. Euclid expanded the term into catoptrics, the science of mirrors. In *Catoptrica* he calculated the origami shapes light traces when bouncing off reflecting surfaces. From this treatise came the one optical law known to almost everyone: The angle of incidence (the angle at which the light strikes the mirror) equals the angle of reflection. Like a ball bounced off concrete, light caroms off a shiny surface at the same angle at which it strikes. Shine your light on a mirror at an angle of 47.35 degrees, and if the mirror is flat the reflection will bounce off at 47.35 degrees. But how did light bounce when it struck a curved reflector, and why did its reflection get so hot?

The Chinese had seen concave mirrors focus sunlight into a "central fire." Noticing the same phenomenon, Greeks used concave mirrors to spark small blazes, including the flame that lit the Olympic torch. To fan the flames and "evoke your astonishment," the mathematician Diocles wrote *On Burning Mirrors*. The lengthy treatise calculated the precise angles at which concave mirrors, reflecting sunlight, would "produce fire in temples and at sacrifices and immolations, so that the fire is clearly seen to burn the sacrificial victims."

Yet the most famous use of light as a fire starter probably never happened. In 214 BCE, Roman troops besieged Syracuse, a walled city on the eastern edge of Sicily. Syracusans fought back with more than the usual slings and arrows. According to Greek lore, the brilliant Archimedes designed long-range catapults, mechanical "scorpions" firing log-sized missiles, and a giant crane that lifted ships out of the water. Legend also says that Archimedes taught soldiers to beam sunlight off giant mirrors or perhaps their shields. Did light, blessed light, really set fire to Roman triremes?

Archimedes' light ray, celebrated since Roman times, has been replicated in recent years. Aiming mirrors or shields, some claimed to have torched model ships, but others barely saw smoke. Lacking definitive proof, hosts of the Discovery Channel's *Mythbusters* made their own test in 2005. The Mythbusters began by gluing three hundred mirrors to a plywood disc so unwieldy it had to be hoisted by a forklift. Aimed at a five-foot-long wooden ship in San Francisco Bay, sunlight flickered and finally homed in. The temperature rose—to above 100° F, then 200 . . . 260 . . . 270 . . . But the heat topped out at 280° F—not enough to burn wood. Myth busted. Enter President Barack Obama, who in 2010, challenged *Mythbusters* to again test "Archimedes' solar ray," this time using "manpower." On a mission from the president, the show's hosts gathered five hundred middle schoolers, gave each a two-by-four-foot mirror, and unleashed their "raypocalypse." But no matter how carefully the students aimed the sunlight, the temperature on the wooden target ship never approached the 410°F needed to ignite its striped sail. Busted again. A few years later, the journal *Optics and Photonics News* analyzed the various tests of Archimedes' rays. The columnist Stephen R. Wilk calculated that that solar heat could easily have done Archimedes' bidding if the mirrors had been precisely aligned. But in the heat of battle, targeting moving ships, in an attack that would have had to come at midday, light would have been far less deadly than a barrage of arrows. Still, the dream (or nightmare) of light as a weapon has endured—in legend, in science fiction, and in plans at the Pentagon.

By the dawn of the Christian era, the lighthouse of Alexandria was renowned as one of the Seven Wonders of the World. And beneath its beam, in about 160 CE, the last of light's early students flared the torch the Greeks had lit.

At his home near Alexandria, within view of the Pharos, Claudius Ptolemy crafted his masterpiece, *The Almagest.* The book's planetary cycles and epicycles promoted the image of the earth-centered universe that would reign until the late Renaissance. But somewhere in Alexandria's vast library, Ptolemy had also read Euclid's *Optica.* After more than four hundred years, this first book of optics needed an update. Euclid had seen light as rays spreading outward in a cone. Images seen across a field were dimmer, Euclid said, because rays, scattering like arrows, flew past them. Ptolemy recognized the error. If light rays missed distant objects, such objects "ought to appear fragmented [like a mosaic] rather than continuous." And stars would be invisible, since rays shot from the eyes, expanding over vast distances, would pass them entirely. Ptolemy had a more subtle view of light. Like "heat in relation to the heater," he wrote, light rays must weaken with distance, rendering far-off objects indistinct.

Ptolemy agreed with the pre-Socratics that light—a "visual flux"—came from the eye. To track its path, he built the first optical tool—the diopter. Using this brass disc etched with 360 degree marks, Ptolemy proved Euclid's axiom "The angle of incidence equals the angle of reflection." He charted the angles in his own origami-like sketches, then began studying refraction. Since the days of Homer, Greeks had noticed how light bent as it passed through glass. This phenomenon had spawned primitive lenses used by artisans and emperors—Nero fashioned an emerald into a monocle. But these first lenses, gritty and irregular, focused light into spots and haloes, or just sent it careening. With his diopter in hand, Ptolemy began calculating the precise angles of refraction that would tame light. His calculations started with an experiment fourth-graders still perform.

Drop a coin into an empty cup. Have a glass of water handy. Now raise the cup to eye level until you can no longer see the coin. Gently pour water in the cup. As the level rises, the coin becomes visible again, its image refracted over the rim. This familiar experiment comes from Book Three of Ptolemy's *Optics.* To calculate the angle of refraction, Ptolemy placed his brass disc in an empty bowl and filled the bowl with water. Lowering his eyes along the disc, he measured refraction at increasingly oblique angles. When seen from just 10 degrees above the water, light was bent by 8 degrees. From 10 degrees higher, the refraction increased to 15 degrees. Measuring every 10 degrees, Ptolemy calculated each refraction. But he used arithmetic instead of trigonometry and his angles were mostly incorrect. A thousand years later, using a similar diopter and the sine function, an Arab scientist would get the math right.

Like Euclid, Diocles, and China's Mohists, Ptolemy also studied convex and concave mirrors. The convex surface made images appear smaller, he noticed, but the concave were trickier, inverting images at a distance, then flipping them right side up when they passed inside the focal point. (Try it with a shaving mirror.) Through ninety-five painstaking theorems, Ptolemy calculated how convex mirrors played with light. But like others before him, he could not resist common folklore. In his companion to *The Almagest*, *The Tetrabiblios*, Ptolemy explained how light from sun and moon affected seasons, weather, and the growth of plants and animals. Yet despite his pseudoscience Ptolemy was the first to bring light into a laboratory. The result was the five books on optics that Ptolemy gave the world. The world ignored them.

"More comprehensive and technically demanding than it was worth," historian A. Mark Smith wrote, Ptolemy's *Optics* "sank into oblivion almost as soon as it was produced." The Pharos beamed on into the Christian era. Ptolemy's *Almagest* ruled astronomy, but few cared enough about his *Optics* to even drop a penny into a cup of water.

Like today's Ph.D. candidates, light's first students had dug ever-narrowing niches. Explaining their discoveries, however, they were far ahead of the rest of humanity. From Aristotle: "Light is a 'nature' inhering in the Translucent when the latter is without determinate boundary. But it is manifest that, when the Translucent is in determinate bodies . . ." From Ptolemy: "Let ABG be the arc of a circle lying on a convex mirror whose center is D, let E be the eye, and let the visible object occupy points H and Z on both sides of the eye . . ." Throughout the first centuries of the Common Era, every learned person knew astronomy and philosophy, geometry and geography, but the inchoate science of optics engaged almost no one. Another thousand years would pass before optics began to rival the wonder that light inspired in the devout. On into the first millennium CE, light remained, if not a piece of God, then surely His holiest garment, one easily donned by those calling themselves the lights of the world.

CHAPTER 3

"The Highest Bliss": The Millennium of Divine Radiance

Sing a song to the heaven-born light that searches far.

—THE RIG VEDA

By the middle of the third century CE, prophets and holy men had swept across Mesopotamia like conquering hordes. Zoroastrian temples blazed with eternal flames but newer faiths, as if borne by desert winds, had blown into the Fertile Crescent between the Tigris and Euphrates Rivers. Mandeans worshipped a King of Light. Mithraists hoped to reach heaven by inhaling light. Then there were assorted cults: Babylonian, Gnostic, and, most recently, followers of a certain "light of the world" who had died on a cross. So one day in 253 CE, when news spread of yet another prophet promoting his own version of paradise, the Mehr-shah invited this "Envoy of Light" to his garden on the banks of the Tigris.

Persians had long equated gardens with paradise, and the Mehr-shah's garden, lush with palms and ferns, encircled by walls keeping the desert at bay, earned the comparison. Invited into the greenery, the goateed prophet who called himself Mani began to speak of his faith. The Mehr-shah cut him short. "In the paradise which you praise," the Mehr-shah asked Mani, "is there such a garden as this garden of mine?" The prophet's answer is a matter of faith, but stories soon spread of the visions he conjured. The Mehr-shah saw God on his throne surrounded by blazing suns. God is light, Mani said, and light is in all of us. The prophet told of how the Kingdom of Darkness had invaded God's Kingdom of Light, and how the war between the two would determine the fate of the earth. As his garden glowed, the Mehr-shah swooned and fell to the ground. When Mani woke him, the Mehr-Shah urged the Envoy of Light to spread his gospel. Within a generation, Mani's "beings of light" could be found from India to Rome.

Throughout the millennium following the birth of Christ, while the theories of the Greek *phusikoi* were neglected or forgotten, light became the symbol of all that was holy. From the smallest cult to the mightiest religion, light converted heretics and convinced the devout that they had seen the face of God. In a world swimming in blood and desperate for meaning, too much was at stake for light to be a "thing." Too many faiths were afoot; too many lights shone where they had never shone before. It was an age when divine radiance burst from clouds and blinded the faithful. Not all believed light to be God, yet all but a few saw it as his self-portrait. Divine radiance inspired prophets with names seemingly out of J. R. R. Tolkien—Kartir, Arius, Zurvan—to battle for the true light. The prize was the human soul, sometimes said to be seeking light, but just as often said to *be* light. Apostles saw the light of heaven. God's messengers did not merely speak of light; they glowed. And when words failed, light emerged as the answer to the enduring question, Does God exist? There in the clouds, at dawn and sunset, was proof, irrefutable, glorious, transcendent.

Ecclesiastical arguments raged as to what kind of light God might be. Created light or eternal light? What the Romans called *lux* (direct light) or what they called *lumen* (reflected light)? Mani and his followers saw God as actual light, but others saw this as heresy. The arguments diffused language itself. At first, divine light was compared to the sun, then to a hundred suns, a thousand, even "numerous suns and full moons in limitless hundreds of thousands." Theologians defined light in phrases as intricate as Aristotle's, writing of "light from light," "a light to lighten," and "in thy light shall we see light." "God is Light," one bishop wrote, "uncreated Light, and it is in the light of that invisible, uncreated Light that the created lights of the world are visible." Like Greek arguments about the "thing you call light," disagreements about holy light dragged on for centuries. Some, however, were content with ancient stories.

In the Hindu *Bhagavad Gita*, the disciple Prince Arjuna asks the god Krishna to reveal his face. "O Lord, master of yoga," Arjuna says, "if you think me strong enough to behold it, show me your immortal Self." Suddenly the skies begin to burn. "If a thousand suns were to rise in the heavens at the same time," the *Bhagavad Gita* says, "the blaze of their light would resemble the splendor of that supreme spirit." Arjuna is "filled with amazement, his hair standing on end in ecstasy." Finally he speaks. "I see you, who are so difficult to behold, shining like a fiery sun blazing in every direction . . . the sun and moon are your eyes, and your mouth is fire; your radiance warms the cosmos." Trembling, Arjuna begs Krishna to dim his light. The skies cloud over, the earth calms, and the lord again assumes "the gentle form of Krishna."

To witness light's power, however, one did not need to ask a god. A glance at the sun sufficed. Centuries would pass before a more nuanced illumination emerged, centuries in which light formed pillars of faith. In all major religions, light's meaning was manifold, encompassing salvation, revelation, the doors of the hereafter, and the essence of the spiritual. "The night is far spent, the day is at hand," the Apostle Paul wrote to the Corinthians. "Let us therefore cast off the works of darkness, and let us put on the armor of light." But long before divine light was crafted into symbol and metaphor, millions believed that light *was* God. Thus had spoken the founder of the Western world's first major religion, the man known as Zoroaster.

Zoroaster was born in brilliance. According to legends dating from northern Persia in 1000 BCE, before his birth Zoroaster's mother blazed with a light that flooded her entire village. Light continued to beguile Zoroaster, who at age thirty had a vision of a man dressed in pure radiance. The vision led the budding prophet to the teachings of the supreme god Ahura Mazda, or "Wise Lord." Zoroaster preached that Ahura Mazda had created the universe out of infinite Light. Beyond this brilliance, however, lurked the Demon of Darkness who could see the Kingdom of Light yet could not penetrate it. But when Ahura Mazda created a second world, a material world, Darkness entered to make the seas salty, fire smoky, deserts dry and bleak. With the help of Zurvan, the god of time, Ahura Mazda fought to save paradise by separating good from evil and light from darkness. Until the Realm of Light was fully restored, however, Zoroastrians were told to nurture the light within, an elusive glow called *xvarenah*. Linked to human semen, *xvarenah* was the radiant fluid of salvation containing wisdom and spirit. Only when Darkness was finally conquered would *xvarenah* "shine continually over the earth . . . And [this light] will be their garment, resplendent, immortal, exempt from old age."

Facts about Zoroaster are sketchy. Much of Zoroastrian scripture was destroyed when Alexander the Great invaded Persia in the fourth century

BCE. Only a dozen or so hymns, called the Gathas, survive. These, said to have been written by Zoroaster, praise "the principles of Truth and Light." "Tell me truly, Ahura," one hymn asks, "What artist made light and darkness?" Light's central role in Zoroastrianism made fire sacred. Believers fed fire temples with incense and animal fat, keeping flames burning night and day. A few such temples still blaze in contemporary Iran, as does the Zoroastrian New Year, Nowruz. Each March 21, despite discouragement from Islamic mullahs, Iranians celebrate Nowruz. Hailing the triumph of light over darkness, celebrants carry torches, leap over bonfires, and gather families around candlelit shrines. Aside from Nowruz, however, Zoroaster's light is another paradise lost. Fewer than 200,000 Zoroastrians can be found worldwide. Fewer still follow the prophet who, long after Alexander's invasion, emerged to update Zoroaster's light. This envoy, apostle, and "best friend of the Lights," was also known as Mani.

From that day in 253 CE when he dazzled the Mehr-shah in his garden along the Tigris, Mani roamed central Asia, building what briefly became the world's dominant faith. A painter as well as preacher and public relations man, Mani encouraged artists and poets to spread his gospel. Its warring Kingdoms of Light and Darkness owed much to Zoroaster, but Mani was a religious chameleon who co-opted pieces of Christianity, Buddhism, and assorted cults. Though Mani called himself an apostle of Jesus, he denied Christ's virgin birth and bloody crucifixion. Mani's growing legions worshipped Jesus the Luminous, who was not nailed to a wooden cross but was crucified on a cross of light. In Mani's world, Light and Darkness battled in past, present, and future. Without the slightest nod to metaphor, Manichees saw a very real war raging between these very real enemies.

Darkness won the first battle, so Mani said, scattering the Kingdom of Light into glowing particles. But the "beloved of the beings of light" returned. The renewed struggle saw demons of Darkness swallowing light whole. To save the precious radiance, the Father of Light conquered dark's demons and made the earth and heavens from their flesh. Still, shards of light remained in human souls and in the monsters who feasted on them, so the Father of Light purified himself into wheels of fire, water, and wind that drew glittering particles toward the moon. Mani's followers also believed that divine radiance filled trees, plants, and certain fruits. Their ceremonies included hymns to their luminous god.

Behold the illuminator of the hearts comes,
the lamp of Light that even brightens those in darkness . . .

Behold he comes, the wise king of the Beings of Light,
who apportions the good gifts . . .

In our modern age, the teachings of Mani seem like the stuff of fantasy novels, yet in an age lit only by sun, moon, and stars, godlight was as visceral as blood. Huddled beneath a spangled darkness each night, liberated each dawn, Manichees felt light penetrate as no material substance could. To see it, to feel it flow into one's soul, was to be reborn. But to those who believed that God was not light, that He had made man in His own image, Mani's doctrine was neither childlike nor hopeful. It was dangerous.

Mani died in captivity in 273 CE after falling to convert another royal skeptic. By the year 300, his creed dominated the pre-Christian world. Roman soldiers believed in this religion of light. So did Alexandrian sages, Babylonian princes, and peasants from the Mediterranean to the Ganges. Simple and seductive, Manichaeism had something to charm and something to offend everyone. The attacks against it were led by the followers of another prophet, the one who had called himself "the light of the world."

In 312 CE, the Roman emperor Constantine, having seen a blazing cross above a battlefield, converted to Christianity. Christians rejoiced. After enduring savage persecution in Jesus' name, they suddenly found their faith accepted, even triumphant. Given all they had suffered, they refused to suffer fools who considered God to be *actual* light. Jews, Zoroastrians, and others whose teachings Mani had co-opted were happy to join in the purges. By 350, Manichean scriptures and Manicheans themselves had been burned in Rome and Alexandria. The persecution continued, relegating Mani's legacy to a single word still in use:

> **Manichaean**, *adj.*, of or relating to a dualistic view of the world, dividing things into either good or evil, light or dark, black or white, involving no shades of gray.

Mani's most vocal critic was the sinner turned saint, Augustine. Before he converted to Christianity in 386, Augustine spent his lusty youth as a disciple of the Manichee faith. His devout Christian mother despaired, but Augustine, brilliant and tormented, found solace in Mani's light. "I thought that you, Lord God and Truth, were like a luminous body of immense size," Augustine wrote in his *Confessions*, "and myself a bit of that body." As he matured, however, Augustine was wracked with guilt about the fleshpots of his past. He sought answers from Manichee elders but when told that salvation came through the light in his own body, he began to doubt. Having studied Greek theories about light, Augustine knew it was not solely divine, let alone part of his soul. Once he found Christianity, he began lashing out at Mani's "lengthy fables." Manichees were "insolent . . . wretched creatures [who] know no light except the kind they see with the eyes of the flesh." Augustine soon replaced Mani's incarnate light with metaphor, starting with Genesis and the three days between "Let there be light" and God's creation of sun and moon. God's first words, he concluded, had created a "spiritual, not a bodily light." For the rest of his life, Augustine used that spiritual light to refute Mani. "The soul of man, although it bears witness of the light, is 'not that light,' " Augustine wrote, "but God the Word is himself 'the true light which illuminates every man coming into the world.' "

Under relentless attack, Mani's Kingdom of Light receded. In its place arose the redeeming radiance of Jesus, the metaphorical light of Yahweh, the splendorous light of Buddha, and the blinding power of Krishna. All these holy figures used light as often as Zoroaster and Mani, yet with broader purpose. Because nothing else so deeply probed the infinity of the human imagination, light was the perfect avatar. Perhaps rivers flowed with milk and honey. Perhaps gates were pearly, cows sacred, the lotus a lovely seat for the Buddha. But light had no shape or form to embellish. It could be anywhere, everywhere, all colors, all powerful, evanescent and ethereal, mysterious and magnificent. In sanctifying it, the world's scriptures dazzled their followers.

Hindus, despite Krishna's blaze in the *Bhagavad Gita*, refused to reduce the godhead to pure light. From the ever-questioning Upanishads:

> The highest bliss can't be described.
> But how should I perceive it?
> Does it shine?
> Or does it radiate?

A thousand years before Genesis was compiled, Hindus rejoiced in "the light that is most excellent." Special reverence was reserved for the radiance that rose each morning over the Ganges. Nearly two dozen hymns of the Rig Veda are dedicated to the dawn and its goddess, Ushas.

> Dawn on us with prosperity, O Ushas, Daughter of the Sky,
> Dawn with great glory, Goddess, Lady of the Light . . .
> We have beheld the brightness of her shining; it spreads and

drives away the darksome monster . . .
In the sky's borders hath she shone in splendor: the Goddess hath thrown off the veil of darkness.
Awakening the world with purple horses, on her well-harnessed chariot Dawn approaches . . .
Bringing forth light, the Wonderful.

Buddhists, too, revered a radiance both metaphorical and blessed. When the Buddha was born, so the sutras say, five lights shone above him. These lights would continue to shine for the Gautama Buddha and all who bore the name. Buddhism's many "enlightened ones" include a Buddha of Boundless Light, a Buddha of Unimpeded Light, and Buddhas of Unopposed Light, of Pure Light, of Incomparable Light, of Unceasing Light . . . Any sufficiently enlightened Buddha, it was said, could use a single tuft of his hair to light the universe. And when he attained Nirvana, his beams outshone sun and moon. One oft-told tale even credited light with bringing Buddhism to China.

Sometime around 75 CE, the emperor Ming was startled to see a radiant human body fly overhead. Soon after, the emperor dreamt of a shining, golden man. Both apparitions, aides told the emperor, were of a holy sage said to have lived in India. The emperor sent a delegation to India to investigate. When they returned with the wisdom of the Buddha, construction began on the first Chinese Buddhist temple. Nearly two thousand years later, Buddhist temples, lit only by candles, still abound with radiance. Buddhist light also fills the sutras. The Sutra of Exalted Sublime Golden Light praises Buddhas as

Suns blazing glory, splendor and renown.
Golden in color, eyes fine as pure, faultless lapis,
They glow with the glitter of pure gold . . .

Buddhist heaven is the Pure Land, devoid of sickness and death, filled with brilliant flowers and roaming Buddhas, each giving off "a hundred thousand rays of light." Buddhist scholars still debate whether the light in their sutras is metaphorical or literal. Contemporary Buddhists, including the Dalai Lama, speak of attaining "clear light," the goal of perfect mental acuity. Others are convinced they will see pure light at death, and some believe that the most devout become light, a "rainbow body" ascending at death, leaving only hair and nails behind. But few agree on the lights of Mount Wutai.

Nestled among the rolling peaks of northeast China, Mount Wutai is one of Four Sacred Mountains of Buddhism in China. Snowcapped and often shrouded in mist, Mount Wutai was once home to hundreds of Buddhist monasteries. Some fifty remain, their golden stupas drawing pilgrims from around the world. Many come to see where noted Buddhas began their journeys to Nirvana, others to meditate in shrines nearly thirteen hundred years old. And a few come to see lights. The sightings began in 679 CE, when monks on Mount Wutai saw a five-colored cloud beside a glowing Buddha. When the monks fell into prayer, the light vanished. Moments later, above a great pool of flowers, another monk saw a multicolored light as bright as the sun. Word soon spread. More lights, including shining wheels and colored mists, greeted pilgrims. Buddhists have been coming to Mount Wutai ever since. Some still report seeing lights in the sky, white or rainbow fireballs, and once in 1999, flickering lights like butterflies filling a monastery. Locals talk of "Buddha lights," but many monks discourage such distraction from the light that should be sought within.

Scholars have a word for "Buddha lights" and all other luminescence seen in a religious context: photisms. Though not as common now as during the first millennium CE, photisms are still reported around the world. Real or imagined, these lights serve a profound purpose, according to the religious scholar Mircea Eliade. "Photisms bring a man out of his profane universe or historical situation, and project him into a universe different in quality, an entirely different world, transcendent and holy . . . The experience of Light radically changes the ontological condition of the subject by opening him to the world of the Spirit." Anyone skeptical about the power of holy light need only study the most life-changing photism in religious history.

On an ordinary morning in about 36 CE, a few years after the crucifixion of Christ, Saul of Tarsus was walking with friends on the road to Damascus, in modern-day Syria. Raised in a pious Jewish family, Saul considered himself "a Hebrew of the Hebrews," eager to persecute the followers of Jesus. Even as he set out for Damascus, the New Testament says, Saul was "yet breathing out threatenings and slaughter against the disciples of the Lord." His zeal and rabbinical studies made Saul the least likely Christian convert. Then, on that ordinary morning, a brilliant light appeared in the clouds above the Damascus road. Artists would later paint the light Saul saw. During the Renaissance, Fra Angelico portrayed it as golden rays beaming from a haloed Christ. The Baroque master Caravaggio painted livid spears cutting through shadows. William Blake sketched yellow sheets enveloping a ghostly Jesus. But when Saul came to describe what he saw, he had only words:

> And it came to pass, that, as I made my journey, and was come nigh unto Damascus about noon, suddenly there shone from heaven a great light round about me. And I fell unto the ground, and heard a voice saying unto me, "Saul, Saul, why persecutest thou me?" And I answered, "Who art thou, Lord?" And he said unto me, "I am Jesus of Nazareth, whom thou persecutest." And they that were with me saw indeed the light, and were afraid; but they heard not the voice of him that spake to me.

Saul, soon to be known as the Apostle Paul, became Christ's most eminent spokesman. Though sickly and possibly epileptic, Paul spent the next three decades roaming the Mediterranean, covering some ten thousand miles preaching the divinity of Christ. Jailed and flogged, stoned and shipwrecked, he converted whole communities, wrote the eloquent epistles found in the New Testament, and reshaped the light of the Bible.

Throughout the Old Testament, biblical scholars note, God is found as often in darkness as in light. On Mount Sinai, He spoke to Moses from the burning bush but then receded into shadows. When Moses, like the disciple in the *Bhagavad Gita*, asked to see God, Yahweh did not blaze forth like "a thousand suns." Instead, He remained in darkness, a voice telling Moses, "There shall no man see me, and live." Mysterious and shadowed, the Old Testament God preferred to "dwell in the thick darkness." Yet God's reluctant radiance freed the Old Testament from exhaustive descriptions of infinite suns, of light boundless, immeasurable, unimpeded . . . In lieu of making light God's self-portrait, the Old Testament made it into metaphor.

When its dark clouds lift, the Old Testament features the richest tribute to light in Western literature. For those who turn to it in need, light offers Sanctuary: "The Lord is my light and my salvation: whom shall I fear?" (Psalm 27:1); Understanding: "I have even heard of thee that the spirit of the gods is in thee, and that light and understanding and excellent wisdom is found in thee" (Daniel 5:14); Omnipotence: "I form the light, and create darkness: I make peace, and create evil: I the Lord do all these things" (Isaiah 45:7); Inspiration: "I shall arise; when I sit in darkness, the Lord shall be a light unto me" (Micah 7:8); Guidance: "Thy word is a lamp unto my feet, and a light unto my path" (Psalms 119:105); Truth: "O send out thy light and thy truth; let them lead me" (Psalms 43:3); and Justice: "The light of the righteous rejoiceth but the lamp of the wicked shall be put out" (Proverbs 13:9).

Saul of Tarsus, when he studied the Old Testament, must have read these poetic tributes, yet when he became Paul the apostle, converted by "a great light round about me," he began to use light in ways the Old Testament had not. In letters to disciples, Paul's light was not the symbol of a God too mighty to be seen but the radiance that had dropped him to his knees on the road to Damascus. "To tell Paul that the Christ he knew was dead was as though a man told him that the sun was dark," one biblical scholar wrote. "He knew otherwise. He felt the power. He had been blinded by the light." Paul, too, used metaphor, yet his light is neither the boundless light of the Buddha, the splendiferous dawn of the Rig Veda, nor the incarnate light of Mani and Zoroaster. Instead Paul shone a narrow, piercing spotlight on salvation. Those who followed reflected his radiance.

Of the four Gospels written after Paul's death, three mention light sparingly. But the fourth, the Gospel of John, hails Jesus as "a shining light," "the true light," and "a light into the world." Only in John does Jesus call himself, twice, "the Light of the World," the second time just before restoring sight to a blind man. Over the next two centuries, as Christianity grew from a cult into a movement and from a movement into the West's dominant religion, John's became the most prominent Gospel, as vital to clergymen as it was to fishermen. Still, schisms arose.

In May of 325 CE, holy men from throughout the Mediterranean gathered in the city of Nicaea near the Sea of Marmara in what is now Turkey. As Christians once under Roman rule, many bore scars—patches over gouged eyes, backs lashed by whips, awkward gaits due to slashed hamstrings. But they had come to Nicaea, invited by Rome's newly converted Christian emperor, to seek reconciliation. In Constantine's imperial palace, not far from the city's ancient walls and amphitheaters, bishops debated, prayed, chanted. No such synod had ever been held, and with Christianity ascendant, there was much to discuss. Baptism and other rites. The proper date of Easter. And above all, whether Christ the Son was as holy as God the Father. For nearly a decade, a clergyman named Arius had been preaching what many considered a heresy. Christ, Arius argued, was not part of some mystical Trinity equating Father, Son, and Holy Ghost. Only God was God "unbegotten . . . eternal . . . without beginning." Jesus was "God, but not true God." Using songs and sermons, Arius had spread this idea throughout the eastern Mediterranean, splitting Christianity in two. What might heal the divide?

Seeking a link between God and Christ, bishops chose light. Just as one candle lights another without losing its glow, one deacon argued, so God's light begat Christ's radiance. Debate continued, but light prevailed. Bishops at Nicaea approved the Nicene Creed, which proclaimed Christ to be "God from God, Light from Light, true God from true God." And when the council adjourned, Arianism was branded heretical, Christ was affirmed as part of the Trinity, and despite further skirmishes, Paul's light was triumphant. Decades later, Augustine, converted by reading Paul's epistle to the Romans, praised what he called the "unchanging and incorporeal light." Augustine's writings used the Latin words *lux* and *lumen* more than four thousand times.

Paul's spotlight returned at the close of the New Testament to offer eternal salvation. The Book of Revelations begins when John—scholars disagree on whether this is the same John who wrote the Gospel—sees seven glowing candles and the shimmering image of Jesus. "His head and his hairs were white like wool, as white as snow; and his eyes were as a flame of fire . . . And his countenance was as the sun shineth in his strength." Detailing the coming apocalypse, Revelations uses light to calm the faithful. After the Second Coming, "There shall be no night there; and they need no candle, neither light of the sun; for the Lord God giveth them light; and they shall reign for ever and ever." Paul's light of salvation became the accepted sign of Christ's presence, a glow that would appear before the holy for the next two thousand years. Joan of Arc saw "brightness [that] comes at the same time as the voice." St. Theresa of Avila saw "not a brilliancy which dazzles, but a delicate whiteness, and a brilliancy infused . . ." And in recent decades, as medical science has brought patients back from the brink of death, many have returned with stories of a light—strong, benevolent, and welcoming. (These are investigated in the Appendix.)

The light of the Old Testament, metaphorical and elusive, still soothes Jews and Christians. The light of the New Testament, focused and redeeming, still blazes in churches and cathedrals, in hearts and souls. Other holy lights kindled during the first millennium remain the truth that millions read into every sunrise, every sunset, and all the glittering stars. Yet this splendor poses a nagging question. What some call photisms might others call hallucinations?

Paul's Damascus light has long challenged skeptics. Carl Jung saw Paul's vision as a "psychogenetic condition" caused by his "fanatical resistance" to Christianity. Contemporary psychologists blame infantile regression, sexual frustration, or psychotropic drugs. Neurologists suspect sunstroke, seizure, or epilepsy. But many consider such speculation irrelevant. The psychologist William James, who catalogued photisms in *The Varieties of Religious Experience*, cautioned against dismissing any vision of light as "hysterics." That such light "should for even a short time show a human being what the high-water mark of his spiritual capacity is," James wrote, "this is what constitutes its importance."

Still, did Zoroaster really see a man dressed in light? Did the Mehr-shah in his garden see Mani's luminous kingdom? Did Paul see a light on the road to Damascus? Skeptics shake their heads; the devout grow more devout. And here light divides humanity into two camps, those content to be enthralled and those who must explain. The source of divine light is a question for the ages, not one that troubled the age of Highest Bliss. Once light had built pillars of faith, centuries would pass before skeptics began to topple them. The Western world calls this period the Dark Ages, a time when scientific curiosity about light surrendered to divinity and dogma. Elsewhere, however, the light of inquiry still burned.

CHAPTER 4

"A Glass like a Glittering Star": Islam's Golden Age

> O God appoint for me light in my heart and light in my tomb and light before me and light behind me; light on my right hand and light on my left; light above me and light below me; light in my sight and light in my perception; light in my countenance and light in my flesh; light in my blood and light in my bones. Increase to me light, and give me light, and appoint for me light, and give me more light, give me more light, give me more light!
>
> —MUSLIM PRAYER

During the centuries that saw Europe fall to its knees and China fortify its walls, when Baghdad was the largest city on earth and scientists bore such names as al-Kindi and Ibn al-Haytham, spires of divine radiance spiked the skies from India to the Iberian peninsula. Each was erected to send out a sound, the call to prayer, but each was also a symbol of light. Historians disagree on whether Alexandria's lighthouse might have been the model, yet every Arabic speaker knows that *manara* means lighthouse, or "place of light." And across desert sands, minarets, the graceful towers named for the word, spread their light.

Minarets were not merely named for light. Each year at the start of Ramadan, a lamp placed atop a minaret each morning signaled the start of fasting. All day the lamp burned, barely seen in the sun. Then at dusk, the single lamp gave way to the boldest displays of light the world had yet seen. As darkness swept across the Islamic empire, celebrants gathered beneath minarets to fire up lamps—ten, twenty, fifty earthen bowls of burning oil. Tied to ropes and hoisted skyward, they shone all night, lighting the festivities below. The lights of Ramadan astonished Western visitors. "They light many lamps on the towers of their mosque," a British traveler wrote from Cairo. "While watching this spectacle we were struck by the sight of towers sparkling with light, each of them lit with numerous lamps at three levels. Thanks to those lights, the city has the splendor of the day."

Throughout the Golden Age of Islam (650–1250 CE) natural light was never scarce. Desert skies blazed with sunlight, while low latitudes saved the faithful from the endless winter nights found further north. Yet the Islamic Empire exalted light as no empire before it. The average Muslim possessed just a few lamps and candles, yet caliphs commanded light by commanding the artisans who harnessed its brilliance. Primitives had worshipped light; Greeks had explained it. Christians, Buddhists, and Hindus endowed their gods with its spirit. But Islam *owned* light and the light it owned projected wealth, eminence, and power. Small wonder that tales from the Arabian Nights turned ordinary lamps, rubbed to reveal genies, into the stuff of dreams. Across the Islamic kingdom, sun, moon, and stars followed carefully charted schedules, but within mosques and palaces, light gave command performances.

Having spread farther and faster than any previous empire, Islam was ever conscious of its fragility. From the mountains of Spain to the Hindu Kush, minarets looked out over borders teeming with enemies. Along with exiled Christians and Jews, caliphs feared rival families, jealous heirs, and the crossfire between Islam's own diverging sects, Shia and Sunni. To remain in command, caliphs had to prove their power. Daggers and scimitars were proof in battle, but to deter intruders, power was projected from every monument, mosque, and palace. Power's most elegant symbol was light—light whose brilliance had lit the world when the prophet Muhammad was born, light that had shone from his father's head that morning, light that enthralled and enchanted.

To enhance light's magic, craftsmen carved quartz—"transparent stone"—into crystalline shapes that sparkled near candles or lamps. Potters mixed metal oxides into clay to create ceramics that nearly glowed. Calligraphers wrote in ink infused with gold, creating illuminated manuscripts such as the Blue Qur'an, whose gilded Arabic text seems backlit on its indigo pages. Islamic palaces filtered light through walls of translucent marble or mother of pearl. No royal courtyard was complete without fountains whose waters splashed light as if it were a toy. Mosaics sparkled in mosques, and in one palace on the Iberian Peninsula, sunlight danced off a basin filled with pure mercury. While a slave tilted the basin, onlookers marveled at the quicksilver patterns swirling through the chamber.

The various dynasties of Islam struggled for revenge and control, yet all agreed that light was to be cherished. Their holy book, sometimes called al-Nur, The Light, told them so. Like the Bible, the Qur'an gives light many meanings:

- "God is the patron of the faithful. He leads them from darkness to the light."
- "He will grant you a double share of His mercy; He will bestow on you a light to walk in."
- "Indeed the man from whom God withholds His light shall find no light at all."

But the Qur'an does not mention light as often as other scriptures. Revealed to a single man, the prophet Muhammad, the noble book was not written by disciples competing to praise the blessed light, the infinite light, the immaculate light . . . Yet among the Qur'an's many suras (chapters) is one devoted solely to light. Sura 24, which is etched into the base of minarets and carved into lamps hung in mosques, reads: "God is the light of the heavens and earth. His Light is like this: there is a niche, and in it a lamp, the lamp inside a glass, a glass like a glittering star, fueled from a blessed olive tree from neither East nor West, whose oil almost gives light even when no fire touches it—light upon light—God guides whoever He will to his Light." Along with memorizing this sura, devout Muslims knew the ninety-nine names for Allah, among which are the Exceedingly Compassionate, the Exceedingly Merciful, the Guardian, the Irresistible, the Light . . .

Throughout Islam's Golden Age, Europe's holy men quarreled about light. At issue was the light said to have radiated from the resurrected Jesus when he stepped from his tomb on Mount Tabor. According to the Gospel of Matthew, "His face shone like the sun and his garments became white as light." Paul, on the road to Damascus, also saw this glowing face, called "the light of Tabor." Bishops in the eastern Mediterranean argued that such light was the tangible link between man and God's "uncreated light." In other words, photisms are real. Bishops in Rome, however, held that those who spoke of seeing a glowing Christ were waxing metaphorical, that even divine faces do not glow. And so light cleaved the Christian world, opening a fissure between East and West. As the East-West split widened, golden mosaics gilded Byzantine churches, but not a single European scholar studied the secular properties of light. That inquiry was left to the civilization spreading farther south.

In 762 CE, Baghdad was founded as a small settlement on the Tigris River. Its walls soon encircled markets, a mosque, and a palace with a green dome that could be seen for miles. And from miles around they came—Muslims, Christians, Jews, pagans . . . Within a century of its founding, Baghdad teemed with more than a million people. Bursting beyond its arched gates, the city called to the entire empire of Islam. Five times a day, the call was to prayer, the muezzin's serpentine tones echoing from minarets. But another call had gone out, a call to the curious: Come to Baghdad if you want to learn. Scholars today are not sure of the site, but somewhere in the overflowing city was the House of Wisdom. Though it had no pools of mercury, no splashing fountains, no translucent quartz or marble, the House of Wisdom preserved the light of the ancient world.

Historians speak of the path "from Alexandria to Baghdad." This was the path Greek learning took to survive, and even thrive, under Islam. Throughout the ninth century CE, deep inside a Baghdad library, black-robed scholars labored with stylus and ink, translating scrolls and codices into Arabic. Works translated included the jewels of Persian poetry, the gems of Indian scholarship, and the full canon of Greek wisdom. Aristotle and Plato headed the list, but scholars also translated tragedy and comedy, astronomy and astrology, Empedocles' four elements, even Diocles' treatise on burning mirrors for "sacrifice and immolations." In little more than a century, linguists in the House of Wisdom translated nearly every text that Greeks, Indians, and Persians had saved from conquest and decay. Only a few such texts were taught in Islamic madrassas, where the curriculum centered on the Qur'an, but the wisest of the million or so in Baghdad added their own wisdom to the house. Astronomers improved Ptolemy's star charts. Healers used the writings of the Greek anatomist Galen to open the world's first hospitals. Mathematicians invented the discipline still known by its Arabic name, algebra. Philosophers tackled Greek sophistry, and a few polymaths studied all of the above, including the thing the Greeks called light.

In the progression of light through succeeding civilizations, Islamic science is the wheel within the wheel. Like the engraved brass discs Arab astronomers perfected to chart the stars, Islamic science was intricate, precise, and quite beautiful. While Europeans fought over holy light, Islamic scientists played with the optical puzzles posed by Aristotle, Euclid, and Ptolemy. They built the world's first observatories and stocked them with astrolabes, sundials, and armillary spheres modeling the movements of the sun and planets. Taking their cues from the Greeks, these Arab students of light repeated familiar questions and devised fresh answers. Some, however, built wheels that were wholly original. The mathematician Ibn Sahl was the first to use trigonometry's sine function to calculate light's refraction when striking water or glass. The philosopher Ibn Sina (known later in the West as Avicenna) posited that light must have a precise speed, for "if the perception of light is due to the emission of some sort of particles by a luminous source, the speed of light must be finite." A dozen or more thinkers spun this Islamic wheel but two in particular were at its hub.

Embedded in the name Abu Yusuf Ya'qub ibn Ishaq al-Kindi is the Arabic word for illumination, *ishaq*. Coincidence, perhaps, but this early Arab philosopher, the son of a wealthy family who came to Baghdad in the mid-800s, could not avoid the mysteries of light. Like most others in its thrall, al-Kindi was an independent scholar, brilliant, tireless, and as Thomas Edison would later say of himself, "interested in everything." The name al-Kindi spread through Islamic learning like a gentle guiding light. His treatises explored medicine, mathematics, ethics, astronomy, music theory, and everyday subjects ranging from swords to perfumes, from jewelry to glittering glass. Some in Baghdad scorned the idea that Arabs could learn anything from other cultures. Al-Kindi disagreed: "We ought not to be embarrassed about appreciating the truth and obtaining it wherever it comes from, even if it comes from races distant and nations different from us. Nothing should be dearer to the seeker of truth than the truth itself."

If ever a single man was a bridge between cultures, it was al-Kindi. His vast works linked Athens' academy with Baghdad's House of Wisdom. On the subject of light al-Kindi picked up where the Greeks had left off, asking whether light came from the eye (as Empedocles thought), the object (Epicurus), or both (Plato). Arguing that light came from the eye, al-Kindi imagined a two-dimensional world where a circle was flattened on a horizontal plane. If the circle alone emitted light, its rays should ricochet in all directions, making the full circle visible. Yet the circle appears to be a line because light from the eye strikes it only on the horizontal. Still, al-Kindi could not entirely refute Aristotle, whose works he introduced to the Arab world. Perhaps light from the eye also met light from each object, creating what the Greek Stoics had called *pneuma*, an ethereal essence the eye perceived. Al-Kindi's successors would call such essence "luminous breath."

Though al-Kindi mostly echoed Greek ideas on light, he changed our understanding of how it travels. Light does not move in parallel rays, he reasoned. If it did, each landscape would appear as the tips of such rays, a montage of points. The French artist Georges Seurat would later make masterpieces out of single dots, but as a model for light, al-Kindi found this "worthy of much derision." Instead, he argued that light radiates. "Everything that has actual existence in the world of elements emits rays in every direction, which fill the whole world." Captivated by this concept, al-Kindi proposed that planets and stars must emit rays that influence all earthly events according to Allah's will. Human beings, he went on, radiate beams of hope, faith, and desire. Al-Kindi's all-encompassing rays also radiated from magnets, mirrors, and fire. With light as his model, al-Kindi wrote, "It is manifest that everything in this world, whether it be substance or accident, produces rays in its own manner like a star."

Because he was interested in everything, al-Kindi did not spend much time studying light. He only came to it after writing other philosophical treatises: *Discourse on the Soul, On Dispelling Sadness, On the Intellect*, and hundreds more. But his work on optics, as *De Aspectibus*, was widely read in the Arab world, and when rendered in Latin during the 1100s, spread throughout Europe. By then, the wheel within the wheel had been given another spin.

In 1009 CE, troops from a rival Muslim faction swept north from Morocco to invade the Iberian Peninsula. Their target was the largest palace in Moorish Spain, the Madinat al-Zahra, just outside the Moorish stronghold in Cordoba. Here light gave its greatest show, reflecting off golden ceilings and marble walls, sparkling in hundreds of fountains, dancing across a pool of mercury. Early in November 1009, troops reached the enormous palace. Storming the gates, they smashed fountains, shattered marble walls, and drained the quicksilver pool. Across what the Arabs called al-Andalus, Islam splintered into emirates clinging to power. Baghdad proved more stable, yet once the House of Wisdom finished its century of translation, the center of Islamic learning shifted to Cairo. There on a hill in the center of the ancient city sat the enormous al-Azhar mosque, its name derived from the Arabic word for "shining." From sunrise to sundown, the mosque's arched halls echoed with recitations of the Qur'an. At the end each full reading came the Daybreak sura: "I seek refuge in the Lord of Daybreak from the mischief of His creation." And while the devout prayed, somewhere on the winding streets below, after a thousand years as the source of scripture, light returned to the laboratory.

Alone in a dark mausoleum, a slight man, gray bearded and wearing a large turban, lays a small rug on the dirt floor. Kneeling, he bows his head to the ground and says prayers to Allah. Then he begins his experiments. He closes all windows, allowing a beam of light to enter through a hole bored in a shutter, "a beam that is fairly wide but not excessively so." The single shaft pierces the darkness, then begins to do the old man's bidding. He places object after object in the beam. He charts each shadow, scribbling in flowing Arabic. He notes how light striking dark objects is absorbed, while light reflecting off white objects brightens the whole room. Through his window, he hears the tumult of the street, the vendors, the muezzin, the cacophony of Cairo. He pays no attention. Rising from his table, he fills jars with water and holds them in the light, calculating angles of refraction. He studies spectra cast on the wall. He polishes pieces of silver until they reflect light at predictable angles. He lifts glasses of water tinted with fruit juice, and notes the spreading colors. He plays with burning mirrors, adding his own calculations. Later he will study eclipses, and the light of moon and stars. And he will think and think before finally beginning to write.

He was born Abu Ali al-hasan ibn al-Hasan but was known among Arabs as Ibn al-Haytham. In his hometown of Basra on the Persian Gulf, he had grown up seeking the truths of the Qur'an. But wearied by theological wrangling, Ibn al-Haytham resolved to pursue only "doctrines whose matter was sensible and whose form was rational." His era, straddling the first and second millennia, was the last in which a scholar might reasonably aspire to know everything, yet Ibn al-Haytham was not content to merely "know." A truth-seeker, he felt, should "make himself an enemy of all that he reads, and, applying his mind to the core and margins of its content, attack it from every side." Bored by his government post in Basra, Ibn al-Haytham quit to pursue his own studies in physics, math, astronomy, cosmology, and meteorology.

Taking up civil engineering, he devised a plan to control the Nile River. In 1010, when he was approaching fifty, Egypt's caliph invited him to put his plan into practice. Stories say that when Ibn al-Haytham arrived in Egypt, the mighty caliph rode on a donkey to meet his new engineer along the Nile, then sent him upriver. Floating past magnificent cities, Ibn al-Haytham realized he had underestimated the Egyptians. If they could build such splendorous palaces, such soaring minarets as he saw in Luxor and Aswan, they could surely control the Nile, if it could be controlled.

Born in modern-day Iraq in 969, Ibn al-Haytham solidified the science of optics. Translated under his Latinized name, Alhacen, his works influenced Kepler, Descartes, Galileo, and Isaac Newton.

Stories also say that Ibn al-Haytham, convinced his plan would fail, feigned madness to get out of his agreement. Thrown in jail, he spent eleven years behind bars with light as his lone diversion. Each day he watched it spread across his cell, watched the sky change, the sun set, the stars emerge. Wondering what light might be and how it behaved, he decided, if ever freed, to spend his remaining years chasing this eternal mystery. Finally released when the caliph died, he moved into a cavernous mausoleum near the al-Azhar mosque at the center of Cairo. There, bent and graying yet unbroken, Ibn al-Haytham began studying light.

Sketches of Ibn al-Haytham, with bulbous turban and flowing beard, present him as a kindly grandfather. But it was his own sketches, diagrams accompanying his work on optics, that would end light's childhood. In seven exhaustive volumes, Ibn al-Haytham examined light more thoroughly than anyone else before Newton. Sensing a "confusion" in Greek optics, he weeded out guesswork, making the first study of light that avoided metaphysics, ether, eidola, etc. His implements were crude—glass cubes and wedges, candles and fire, a copper pot, a wooden stylus, and a block of glass "eight digits long, four digits high, and four digits wide." Aside from digits, his chief unit of measurement was the barley grain. Science was his method but never his master. His treatises all end: "Thanks be to God, Lord of the Universe, and blessings be upon His Prophet Muhammad and all his kin." Yet Ibn al-Haytham's experimental acumen presaged that of Galileo.

Turning his mausoleum into an optics lab, this avuncular man manipulated, studied, and played with light. Ibn al-Haytham began with the oldest question about light—its source. Eye or object? His lengthy work opens: "We find that when our sight fixes upon very strong light-sources it will suffer intense pain and impairment from them, for when an observer looks at the body of the sun, he cannot do so properly because his vision will suffer from its light." If the sun hurts the eyes, how could light come from those eyes? And if light poured *out of* the eye, how could it fill the vast space between an observer and the stars yet leave the eye unchanged? Such an idea, Ibn al-Haytham wrote, was "quite impossible and quite absurd." Fourteen hundred years after it began, the debate—Eye vs. Object—was over. Some would cling to their opinions but none who read Ibn al-Haytham could doubt his logic.

Few of light's tricks escaped him. Light, he posited, was not a single cone of radiance, as Ptolemy had said. Light formed pyramids, infinite in number, filling all visible space. Vision occurred wherever the eye met a pyramid of light. Each pyramid, intercepted at a different point a few digits away, created a different image of the same object. To explain how the eye worked, Ibn al-Haytham made the first anatomically correct diagrams depicting human vision from the eye to the brain. He used these to explain color and depth perception, peripheral vision, the optic nerve, and how easily something so wondrous as the eye can be tricked by something so ephemeral as light. All vision, he noted, was refracted through the eye's "crystalline humor," and interpreted by the brain. Yet light remained the source of sight, of perception, even of beauty. "For light creates beauty, which is why the sun, moon, and stars will appear beautiful." Shadows also created beauty, "for in many visible forms there are blemishes and tiny pores that render them ugly . . . but when they are in shadow or in weak light, those blemishes and wrinkles will disappear, so their beauty is apprehended."

Ibn al-Haytham's lone optical invention was the camera obscura. The Chinese Mohists were the first to notice how a hole in a wall streams outdoor light onto an indoor surface, upending the image. Others had played with pinhole apertures, transfixed by the inverted pictures they projected onto walls. But Ibn al-Haytham made the first fully operative light box. A dark room. A small box with a pinhole. Five candles flickering. And inside the box, all five flames were projected on the far side, inverted in perfect proportion. Others might merely have gazed, but Ibn al-Haytham stretched straight edges and copper rods from source to image. Measuring, calculating, he came away convinced that light did not mingle, nor did its rays cross. But was there a "least light," its rays diminished until an object was blacked out of his light box? Through smaller and smaller apertures, the search began. Centuries later, Isaac Newton, having read Ibn al-Haytham in Latin, renewed his quest for "least light."

With his wordy and exhaustive prose, Ibn al-Haytham would have made a good patent lawyer. His description of a basic refraction apparatus, a simple cylinder with holes on opposite sides, runs on for three pages. Still, his conclusions about refraction have never been doubted. Ptolemy had jumpstarted the study of refraction with his brass disc dipped in water. Ibn al-Haytham went further. The denser the material, he observed, the more it bends light. But light is refracted only when it strikes water or glass at an angle. Shine a perpendicular light at a pool and it will hit the bottom at a right angle.

Ibn al-Haytham's refraction experiments seemed simple enough to try at home. I took a pizza pan, hammered nail holes on opposite sides of the rim, then set it in the empty kitchen sink lit by the morning sun. I aligned the sunlight so that it shone through one hole, making a bright spot on the lower rim at the bottom of the sink. Then I turned on the tap. No sooner had water risen above the rim than the white pinpoint began to move, creeping down the pan, toward the perpendicular. By the time the sink was full, the sun's rays were refracted at least five barley grains, about a quarter inch. And when I yanked out the plug, the water drained and the pinpoint edged back, grain by grain. But testing Ibn al-Haytham's observation about perpendicular rays was harder. To capture perpendicular sunlight, you need to live between the Tropic of Cancer and the Tropic of Capricorn, and even there the rays are perpendicular only twice a year. So I bought the "least light" I could afford—one not even the richest caliph in Baghdad or Cairo possessed. Using a five-megawatt laser pointer, I beamed a red ray straight down through my pizza pan. No matter how high the water rose in my sink, the perpendicular light did not refract. Again, Ibn al-Haytham was correct.

Until recently, Western scientists dismissed Islamic optics, claiming it had merely kept Greek science "in cold storage." But thanks to clearer thinking and deeper research, we now know better. Taking the best of Ptolemy, of Aristotle, of Euclid, adding experiments, subtracting speculation, al-Kindi, Ibn al-Haytham, and their disciples created a single, encyclopedic volume of the best thinking about light. Their work sent optics into the Middle Ages with a firm footing. "Without the theoretical groundwork laid by Ibn al-Haytham . . ." A. Mark Smith wrote, "the revolution in optics inaugurated by Kepler and completed by Newton would have been, if not inconceivable, at least difficult to imagine."

Ibn al-Haytham died in Cairo in 1039. In the decades that followed, minarets blazed throughout Ramadan, and light, whether refracted in quartz or reflected in fountains, put on a show. Then, in the 1100s, Islamic tolerance for science, for experiment, for tolerance itself began to wane. Emerging thinkers shunned those still studying the Greeks. One widely read tome, *The Incoherence of the Philosophers*, suggests the growing disdain for philosophical discussions. "May God protect us from useless knowledge," the tome's author, al-Ghazali, wrote. In lieu of experiment, al-Ghazali relied on "the effect of a light which God Most High cast into my breast; and that light is the key to most knowledge." An Arab historian dealt the final blow to the Islamic study of light. "The problems of physics," Ibn Khaldun wrote, "are of no importance for us in our religious affairs or our livelihoods. Therefore, we must leave them alone."

Late in the twelfth century, shortly before Mongol hordes sacked Baghdad and burned its enormous library, a Persian sage tried to weave the unraveling strands of Islamic thought. Known as Suhrawardi, he promoted a philosophy that blended the light-based god of Zoroaster with the emerging mysticism of the Sufis, adding a little Greek and Egyptian philosophy for good measure. Writing sometimes in Persian, sometimes in Arabic, this *Sheika al-Ishraq* ("Master of Illumination") traveled throughout the Near East before settling in Aleppo, in modern-day Syria. Suhrawardi was no scientist. Favoring a strident asceticism, he claimed that no one could understand his major work, *The Philosophy of Illumination*, without fasting for forty days. Having done so, he had seen "beings of light whom Hermes and Plato contemplated, and the heavenly radiation, sources of the light of glory and the kingdom of light which Zoroaster proclaimed." Like Zoroaster and Mani, Suhrawardi saw God as pure light reflected in the human soul. Each of us, he wrote, has vague memories of our luminous origin and each night when our world is in shadow, we yearn for our first abode, light. Light came from neither eye nor object, Suhrawardi argued, but from the soul. A soul purified by fasting would see fifteen types of light, including light soothing like "warm water," a light "of extreme grace and pleasure," a light "more intense than the light of the sun," a light "that gives birth to the self."

With the Crusades under way, the Islamic world had little patience for such mysticism. The Egyptian sultan Saladin had reclaimed Aleppo from Christian invaders, but the rest of Islam remained under siege. In 1208, Suhrawardi was charged with heresy, jailed, and executed. Some say the Master of Illumination was killed by the sword, others that he was strangled or thrown from a fortress wall. Martyrdom would later make his philosophy widely studied in Islamic circles. Sufis enchanted by its eulogies would entomb their saints in marble domes whose soft whiteness symbolized Allah's light. Inspired by Sufi scripture, the Mughal Empire of India would also revere marble's luminescence, creating among other masterpieces, the Taj Mahal. Yet *The Philosophy of Illumination* was never translated into Latin.

CHAPTER 5

"Bright Is the Noble Edifice": Paradise in the Middle Ages

Watch the dust grains moving
In the light near the window.
Their dance is our dance.

—RUMI

Abbot Suger was a wisp of a man, kind and industrious, with a passion for all that sparkled. Unlike his ascetic rival, the abbot felt monks need not suffer endlessly, nor shun finery in contemplation of the divine. Monks in Suger's abbey of Saint-Denis had wooden choir stalls instead of the usual marble that chilled the bones during icy Parisian winters. The abbot's own cell, though lit only by candles, included an occasional cloth or curtain. Yet what truly shocked the abbot's rival was the church's worldly riches—gold and pearl and precious stones. Abbot Suger explained that it was not wealth he sought in gems, but a celebration of God's own light.

Abbot Suger had been in charge of Saint-Denis for just two years when he began planning a wholesale reconstruction of the church. The ancient edifice suffered "grave inconveniences." It was cramped, even "moldering away." Columns were shaky, walls were cracked, and few felt the gloomy church to be worthy of its namesake, the patron saint of France. By 1100, the church at Saint-Denis on the northern edge of Paris had been the spiritual seat of the French monarchy for nearly five hundred years. Kings were crowned there, buried there. Legend said that back when France was just a handful of warring fiefdoms, Christ had appeared with a host of angels to consecrate the church. A nail from Jesus' cross and several thorns from his crown were kept in the church's reliquary. Such treasures, the abbot argued, deserved a gilded frame, so in 1124, as France worried about another war with England, as another Crusade was mounted, Abbot Suger began dreaming.

Had it been left to his rival, Bernard of Clairvaux, the rebuilt church would have been like others in France, stout edifices with bare walls and cloisters that sealed out heaven and earth. Devotion required discipline, Bernard argued. Rejecting all finery, especially within the confines of a church, Bernard and his ascetic order, the Cistercians, had "for the sake of Christ, deemed as dung whatever shines with beauty." But Abbot Suger thought a church should hold heaven in its vaults.

The planning took more than a dozen years, but by 1137, having consulted with architects and clerics, and with Louis VII to procure funding, the abbot was ready. He assembled a team of masons, stonecutters, and other craftsmen, then lay awake at night fretting about details. Dressed in his mud-brown cassock, the diminutive man roamed forests searching for statuesque oaks and chestnuts, hailing the miracle of finding the exact numbers needed for scaffolds and roofing. Equally miraculous was the discovery of a "wonderful quarry . . . yielding very strong stone such as in quality and quantity had never been found in these regions." Peering into the quarry pit, the abbot praised God to see how "noble and common folk alike would tie their arms, chests, and shoulders to the ropes, and, acting as draft animals, draw the columns up." Just seven years after the rebuilding began, the church apse, choir, and rose windows were finished. In the spring of 1144, invitations went out to the most powerful men in France and England. On June 11, a pompous procession of holy men, led by Louis VII and Queen Eleanor of Aquitaine, entered the church at Saint-Denis. As they gazed upward, their mouths fell open and they became children again.

In the centuries after Islamic science crafted light's wheel within a wheel, Europe spun divine radiance into spheres beyond spheres. The goal was lofty: to recreate paradise on earth. Throughout the first millennium, whether labeled heaven, Nirvana, or the Pure Land, paradise had been a promise. Scripture and folklore spoke of gardens and haloes, thrones and gates, each seen only after departing this dreary world. Medieval sages had a different perspective. Paradise, they reasoned, was too glorious to consist of gates or thrones and too important to remain invisible. It had to be made of something precious, something perfect, something you could see from here.

As the era's dominant thinker, Thomas Aquinas, wrote, "If, then, there exists a still higher heaven, it must be wholly luminous." The commoner felt the same. Gazing into the night sky, medieval men and women did not see a starry void but a darkness they assumed to be the earth's shadow. Beyond the blackness, as the stars suggested, must surely blaze a radiant paradise, what Aquinas called "the brightness of glory which differs from mere natural brightness." At the height of the Middle Ages, Dante Alighieri would imagine a journey through this "Paradiso" whose light was "so living that it trembles in your sight." But by the time Dante forged heaven's radiance into verse, it was already beaming through stone and glass. Today, nearly nine hundred years after its creation, this first Gothic light still shines.

Here in the church of Saint-Denis is the light Abbot Suger bestowed upon the Middle Ages. Step inside the door, crane your neck, and open your eyes to wonder. Saint-Denis is half the size of Paris's Notre Dame, yet the latter, for all the fame of its facade, is gloomy and cavernous inside. Saint-Denis, by contrast, uses its daring design to yield a light worthy of eternity. If Genesis 1:3 were somehow sculpted in granite, these stones would be its edict: "Let there be Light!" If light could be played by an orchestra, this would be its overture. This—the first Gothic cathedral. Roaming in the warm light, I gape at the glittering walls and the two enormous rose windows in greens, blues, and reds. And there in one glass panel, kneeling at the feet of the Virgin, is Abbot Suger's likeness—in green robe, eyes uplifted. Across the nave, Suger's own words are carved in stone:

FOR BRIGHT IS THAT WHICH IS BRIGHTLY COUPLED WITH
THE BRIGHT.
AND BRIGHT IS THE NOBLE EDIFICE WHICH IS PERVADED
BY THE NEW LIGHT.

Stay awhile and study the light. See how the colored windows dim when clouds move in, then suddenly blaze when the sky clears. Each morning, east-facing windows catch fire while those on the opposite wall remain dull and gray. Then, as if Abbot Suger had made the sun a light switch, mid-afternoon radiance fills the west windows, letting them burn until dusk. Light does this in every home, of course, but in Saint-Denis it also tells stories, paints pictures, and makes a coloring book of the Bible. Visitors, visibly humbled, walk with hands clasped behind, speaking in hushed tones. Because most are French, Saint-Denis is unlikely to be the first Gothic cathedral they have visited. But imagine if it were. Imagine entering it from the mud and misery of the Middle Ages. What promise would you not make to live in this light? Bathed in its glory, you might doubt your chance of reaching paradise, yet the goal would be as clear as the light itself.

Despite Saint-Denis's radiance, however, few contemporary visitors fall to their knees. Conversions on the spot rarely occur. The light of Saint-Denis today is the same that streamed through its windows on the day of its convocation in 1144, yet the light within its visitors is diffused by science and skepticism. To understand how the Middle Ages *felt* rather than saw divine radiance, one must consider the "metaphysics of light" that pervaded those centuries of faith.

The gray summers and long winters of northern Europe have always made sunlight a blessing here, yet Abbot Suger saw the divine in the slightest reflection. Jewels, gold, even ordinary mirrors were holy, the abbot said, because they reflected the light of God. The link between ordinary sparkles and divine light came from a philosopher whom modern scholars give an odd name—Pseudo-Dionysius. This sixth-century mystic falsely claimed to be the same Dionysius converted by the Apostle Paul in the book of Acts. Pseudo-Dionysius actually lived centuries after Paul, yet those centuries were lost in the confusion of the so-called Dark Ages. By 1100, Abbot Suger and others in France were convinced that Pseudo-Dionysius was the same Dionysius/Denis as their patron Saint Denis. The confusion helped Pseudo-Dionysius shape the light of the Middle Ages.

Pseudo-Dionysius' writings are saturated with light, aka "Divine Light," "the Divine Rays," and "The First Gift and the First Light." Pseudo-Dionysius believed that the God who hid his face from Moses also shied from showing his full brilliance to the world. Instead, the "One Source of Light" spread His radiance in flashes, gleams, and twinkles. "Every creature, visible or invisible, is a light brought into being by the Father of the lights," Pseudo-Dionysius wrote in *The Celestial Hierarchy.* "This stone or that piece of wood is a light to me . . . for I perceive that it is good and beautiful." If ordinary objects suggested God, then gems and shiny metals openly sang His praises. Pseudo-Dionysius, in words that spoke to Abbot Suger, wrote of the "unpolluted radiance of gold," and "the heavenly glow of silver." Given the power of light to summon the divine, Abbot Suger reasoned, church windows should be expansive, vaults lofty, jewels abundant. The abbot carved his rationale into a stone in Saint-Denis that reads, "The dull mind rises to the truth through material things." Fortunately for dull minds, medieval architects were learning to fashion earthly materials into portals of radiance.

Back in the mid-sixth century, the architects of the Hagia Sophia in Constantinople (modern-day Istanbul) were the first to bring the light of heaven indoors. The dome of the Hagia Sophia, soaring to nearly two hundred feet, is encircled by forty windows that send beams streaking like swords through the hazy interior. Dozens of other windows flood the church's core with sunlight, yet corridors and apses remain shadowed. In the five centuries that followed this Byzantine masterpiece, mosques and Romanesque churches had as many windows as possible, yet each window weakened its wall, increasing the risk of collapse. Windows in Spain's cathedral at Santiago de Compostela, begun in 1075, comprise just 25 percent of the wall surface. Abbot Suger wanted more windows, more light.

Working with architects whose names remain unknown, the abbot adopted three structural improvements that would usher in the age of Gothic. The ribbed vault featured crisscrossed arches that channeled gravitational stress into adjacent columns. The pointed Gothic arch proved stronger than the Roman arch, enabling windows to be stretched as if on a rack. And the flying buttress, whose graceful exterior arms surround the apse at Saint-Denis, supported the soaring stones, letting vaults rise still higher, enabling glass to serve as wall. "The arrangement," wrote the historian Henry Adams, "is almost too clever for gravity." And the arrangement delighted Abbot Suger. His new church's choir was a "crown of light." His "most luminous windows" filled 78 percent of Saint-Denis' wall surface, giving light a fresh new purpose.

Suddenly divine radiance, no longer at the whim of the clouds, became a daily presence. The Islamic world had made light the plaything of caliphs and princes, but medieval Europe offered its wonders to anyone within walking distance of the great stone structures that began to rise across northern France: Rouen (1145), Senlis (1153), Notre Dame de Paris (1163), Strasbourg (1176). Abbot Suger was no architect or saint. He never wrote a single theological treatise. Yet his faith in all that sparkled exalted light for the common man.

The abbot's rival seethed. "The church sparkles and gleams on all sides while its poor huddle in need," Bernard of Clairvaux wrote of Saint-Denis. "Its stones are gilded, while its children go unclad." Shortly after Saint-Denis's consecration, Bernard's Cistercian order banned color and imagery from its church windows. Abbot Suger answered in a memoir justifying the building of Saint-Denis. "We, most miserable men," he wrote, "should deem it worth our effort to cover the most sacred ashes . . . with the most precious materials we possibly can." Defending the church's sparkling treasures,

Abbot Suger invoked higher authorities—the holy martyrs enshrined in the church. "It was as if they wished to tell us through their own mouths, 'Whether you wish it or not, we want only the very best.' "

In the fall of 1150, as Saint-Denis' nave was being converted to Gothic splendor, Abbot Suger contracted malaria. As Christmas approached, he lay on his deathbed. His prayers to live through the holiday were answered, but he died in January 1151. By then, Gothic light was triumphant. "You paint for yourself the heavens and the gods of the heavens in [your] church," one cleric had written to the abbot. Other men of faith conceived more enduring tributes.

Along with Louis VII and his queen, the Saint-Denis convocation in 1144 had attracted seventeen bishops from France and beyond. Each went home with plans to capture light in his own apse, his own stained glass. On through the twelfth century, the Gothic slowly rose, not just in France but also in England, where cathedral construction was begun in Wells, Lincoln, and Salisbury. With time and technology, stained glass brightened, vaults soared, and the abbot's holy light spread. Then the older churches, damaged by fire, were rebuilt in high Gothic. Canterbury (1174), Chartres (1194), Rheims (1211) . . . Scholars now believe that more stone was quarried in Northern Europe during the Middle Ages than was used in Egypt's pyramids. This building frenzy took place in times that were neither the best nor the worst. Despite a prosperity that doubled Europe's population, life expectancy stayed at forty-five years. The typical house was a hovel whose mud walls had perhaps a single small window covered with milky glass and often shuttered. The bulk of the populace was illiterate, but light spoke to priest and peasant alike. Anyone could step into a cathedral to drink in its radiance.

The label "Gothic" was never used during the Middle Ages. Abbot Suger called his style "modern." Only during the Renaissance was "Gothic" applied—as an insult suggesting barbaric Goths. The Protestant Reformation regarded these cathedrals as grotesque examples of excess. The French Revolution targeted them for "dechristianization" that sent vandals tearing through Notre Dame and Saint-Denis. Not until the mid-nineteenth century were Gothic cathedrals again revered. Victor Hugo called each a "symphony in stone." The sculptor Auguste Rodin praised these "vast poems" through which he walked "with exquisite joy." And today, even the most skeptical, stepping into the light of Rheims or Saint-Denis, risk feeling more than they expected. Two centuries after Napoleon visited Chartres, his observation remains true: "Chartres is no place for an atheist." For even in our own age, when light bursts from every hand-held screen, Gothic radiance recalls the goal of Abbot Suger: to "brighten the minds, so that they may travel, through the true lights, to the True Light."

Cloistered in gloomy monasteries, shivering in cells lit by a single candle, monks lived for light. Its return each morning brought joy; its daytime flood nurtured the spirit, and its evening departure caused even the most faithful to tremble. The moon and stars offered slim consolation. God had to be in their details—somewhere—but only the surrender of night could ease the soul. As the wellspring of medieval faith, light inspired questions unasked for a millennium. Was light pure spirit or some sort of matter? Was all light divine, as Pseudo-Dionysius declared, or were glints and reflections just . . . light? The answers did more than add to the lengthening list of theories. Building on Greek and Arab studies, scholastic monks fused holy and secular light, bringing optics out of the Dark Ages. The works of Ibn al-Haytham, translated under his Latinized name, Alhacen, were widely read in the upstart universities of Paris and Oxford, Bologna and Salamanca. Alhacen had been a Muslim, of course, yet despite the ongoing Crusades, Christians who studied him sensed a kindred spirit. And throughout the thirteenth century, light was studied both through a biblical lens and through lenses concave and convex.

Robert Grosseteste was born in 1168 to a poor peasant family in Suffolk, a land of rolling pastures along England's North Sea coast. As a boy he demonstrated a quick mind that earned him admission to Oxford and to later study in Paris, where he may have visited Saint-Denis. Though now known only to scholars of optics or medieval theology, Grosseteste was among the most formidable intellects of his age. A student of music and astronomy, astrology and Aristotle, Grosseteste wrote on comets, the sun, air, truth, and free will, but he was especially captivated by light. "Physical light," he wrote, "is the best, the most delectable, the most beautiful of all the bodies that exist." Gazing at light, this somber, bearded cleric saw not just rays but an eminence responsible for intelligence, the soul, the universe itself.

Like Alhacen, Grosseteste came late to the study of light, taking it up in his sixties after he became Bishop of Lincoln in 1235. Like Abbot Suger, Bishop Grosseteste believed God's light to be revealed in the slightest sparkle. And like Augustine, he was intrigued by those three biblical days between God's "Fiat lux" and the creation of sun and moon. In his brief treatise *De*

Luce ("On Light"), Grosseteste called the light of Genesis "the first bodily form." Michelangelo would later paint this light in the Sistine Chapel, his God separating billows of brightness from shadow, yet Grosseteste had read of al-Kindi's light radiating in all directions. "Fiat lux," Grosseteste believed, had created a mere pinpoint, a sort of cosmic pilot light. Begat by God, the pinpoint spread ever outward, not as a cone but as a sphere. Light's inner core remained dense, its outer more "rarified," creating larger spheres, which in turn expanded "until the nine heavenly spheres were completely actualized." The highest sphere soon created fire, which expanded to produce air. Air then drew together and, "expanding its outer parts, produced water and earth." Thus God's First Light gave birth to everything.

If light created the universe, then every object must be "some kind of light." Six centuries would pass before the Scottish physicist James Clerk Maxwell proved light to be electromagnetic energy. A generation later, Einstein would show energy to be equivalent to matter. But in 1235, the bishop of Lincoln was onto something. Grosseteste even predicted a time when light would not merely beam into cathedrals but be magnified so that "we may make things a very long distance off appear as if placed very close, and large near things appear very small." Within fifty years of his prediction, the first eyeglasses were ground in Venice. By then, Grosseteste's tomb basked in the light of Lincoln Cathedral.

Grosseteste's *De Luce* was added to the shelf of works on light, yet it was another Oxford don who made the Middle Ages' greatest contribution to optics. Like Grosseteste, Roger Bacon was a friar—cowled, bearded, and intent on explaining the natural world. Though trained in theology, Bacon fell under light's thrall and spent two decades studying optics. He also "spent more than 2,000 pounds on secret books and various experiments and languages and instruments of mathematical tables, etc." The result was *Opus Majus*, a thick tome now considered to be the seed of modern science. Written in a single year, 1267, Bacon's opus spanned the breadth of human knowledge from theology to alchemy, from languages to ethics, from math to the behavior of light. Of optics, Bacon wrote, "It is possible that some other science may be more useful, but no other science has so much sweetness and beauty of utility." In light Bacon saw a hierarchy of righteousness. Direct light, he said, was akin to a direct contact with God. Light refracted through clouds suggested the realm of angels, whereas light reflected by worldly objects symbolized the faith of mere mankind.

Robert Grosseteste had seen light as corporeal, yet Roger Bacon preferred to call light "a species . . . a propagation multiplied through the different parts of the medium." As a species, Bacon maintained, light must have a finite speed, for "it is impossible that there should be an instant without time, just as there cannot be a point without a line." But light's speed had to be blinding, "since any one has experience that he himself does not perceive the time in which light travels from east to west." Bishop Grosseteste had predicted that light would someday be magnified; Roger Bacon saw further. Praising "the wonders of refracted vision," Bacon envisioned a day when "from an incredible distance we might read the smallest letters and number grains of dust and sand . . . So also we might cause the sun, moon, and stars in appearance to descend here below, and similarly to appear above the heads of our enemies, and we might cause many similar phenomena, so that the mind of a man ignorant of the truth could not endure them."

Such wonders lay far in the future, however. Meanwhile, the cathedrals rose, taller, brighter, more celestial. Burgos (1221), Beauvais (1226), Cologne (1248), Orléans (1278). On each site, hundreds of workers—masons, carpenters, blacksmiths, plasterers, roofers—climbed wooden scaffolding or slogged in muck. Toiling sunrise to sunset, they strained at enormous stones and shaped molten glass, all to capture light. Each edifice would take decades—sometimes centuries—to finish. And while all that glass was tinted, all those stones hoisted, another scholarly monk labored to fuse light's diverging paths.

Fellow students in his Dominican order in Naples called Thomas Aquinas "that dumb ox." Ponderous and obese, Aquinas nonetheless impressed his teachers. "The dumb ox will yet make his lowing heard to the uttermost ends of the earth," one predicted. Aquinas, born in 1225, devoted his life to inquiry. Roaming from Paris to Italy and back, he studied, prayed, and forged his thinking into his masterpiece, the *Summa Theologica*. Blending reason with religion, this seminal work unfolds as debate: "Objection 2 . . . Reply to objection 2 . . . On the contrary . . . I argue that . . ." Reason, Aquinas argued, could prove God's existence, biblical truths, even the nature of light. In the *Summa Theologica*, Aquinas summed up medieval inquiry into divine radiance, asking:

- Whether light is a body?
- Whether light is a quality?
- Whether the production of light is fittingly assigned to the first day?

Judging whether light is a body or a quality, Aquinas gave three reasons why the Gothic light streaming through glass and stone could not be mere matter: 1) No object could be in two places at once, yet light was everywhere at the same time. 2) No object could move instantaneously, yet "as soon as the sun is at the horizon, the whole hemisphere is illuminated from end to end." 3) All objects decayed with time, yet light remained uncorrupted. Light, Aquinas concluded, is "an active quality consequent on the substantial form of the sun or of another body that is of itself luminous." Aquinas also found light "fittingly assigned to the first day" of creation, for "day cannot be unless light exists, which was made therefore on the first day."

Lest anyone doubt that light is divine, Aquinas reasoned away all skepticism. Before Adam and Eve, he wrote, the universe was pure light. Their fall from grace dimmed the ambient radiance, but upon Christ's return, light will illuminate a cosmos of "water as crystal, the air as heaven, fire as the lights of heaven." And answering ascetic monks critical of Gothic splendor, Aquinas argued that "the created light is necessary to see the essence of God." Therefore, gold, jewels, and luminous windows should be in every cathedral.

Thomas Aquinas died in 1274, en route to the Second Council of Lyon. He never reached the Lyon cathedral, still under construction. By the start of the next century, hundreds more Gothic marvels were shining on the faithful from Sweden to Spain, but it took a poet to burn this incandescence into the medieval mind.

Gothic light never penetrated the churches and cathedrals of Italy. Built of brick instead of stone, these structures had few windows, most made of clear, not stained, glass. Throughout Italy, heaven kept its distance. It finally appeared in the early 1300s in the vision of Dante.

The year of Dante's birth, 1265, matters less than the year he first saw his beloved Beatrice, 1274. Scholars suspect he never spoke to the eight-year-old girl, yet he never forgot her. By the time she died, in her twenties, Dante was a rising Florentine politician. He somehow found time to read widely, from Aristotle to Aquinas, from Alhacen to Augustine. Then, in 1301, caught on the wrong side of political feuds, he was exiled from Florence, never to return. For the next two decades Dante roamed Italy, dreaming of Beatrice, of revenge, and, finally, of paradise. He would spend eight years crafting his *Divine Comedy* in which he charted his imagined journey, with Beatrice and the poet Virgil, from the depths of hell to the heights of heaven.

Dante never set foot in a Gothic cathedral, but he knew both the physics and metaphysics of medieval light. Scholars argue that the poet was well-read in the science he called *perspectiva*. Quotes from Dante's poem "Convivia" echo Roger Bacon's optics, and in the *Paradiso*, Dante suggests Alhacen's quest for "least light": "There must be a limit where thickness does not allow the light to pass . . ." A few lines later, Beatrice proposes an optical experiment. Take three mirrors. Set two at an equal distance and a third farther back. Place a candle in front and "although the light seen farthest off / seems smaller in its size, still you will observe / that it must shine with equal brightness."

Dante also knew light's most basic reflective property. One afternoon in Purgatory, he and Virgil are walking with the sun in their eyes. Dante lifts his hands . . .

> to limit some of that excessive splendor
> as when a ray of light, from water or
> a mirror, leaps in the opposed direction
> and rises at an angle equal to
> its angle of descent, and to each side
> the distance from the vertical is equal
> as science and experiment have shown.

Dante envisioned Hell as a place "where every light is muted" and a Purgatory so gloomy that it cast no shadows. Through these realms, Dante trudged verse after verse until his fervid imagination took him to Paradise. There he found lights to rival the blazing Hindu god Krishna and the radiance of the Buddhist Pure Land.

Paradiso begins in the Garden of Eden, where Dante and Beatrice stare at the sun, "the lantern of the world." Peering into its fire, the poet begins his incandescent climb. In heaven's first sphere, Dante dotes on the moon, then suddenly finds himself in a cloud that shines "like a diamond struck by sunlight." His ascent continues to the Sphere of Mercury, where twin lights dance "like the fastest-flying sparks." Next, arriving in the Sphere of Venus, Dante and Beatrice are greeted by lights within lights, "as within a spark one sees a flame." Lights swirl, wheel, and burst into song. From within this radiance comes a voice: "We are all ready at your pleasure / so that you may receive delight from us." Following this light into the Sphere of the Sun, Dante sees

> many living lights of blinding brightness
> make of us a center and of themselves a crown
> their voices even sweeter than the radiance of their faces.

The luminous crown wheels three times around the poet before he recognizes in it the spirits of saints and sages. Among them are the philosopher whose theories of light inspired the cathedrals, Pseudo-Dionysius, and the dumb ox whose *Summa Theologica* outlined the nine spheres of Paradise that Dante is traversing. A second, brighter crown appears, then more and more and still more lights swirling.

As Dante rises, sphere by sphere, the lights of Paradise promenade. They sing. They form a spangled cross. In the sixth circle, the Sphere of Jupiter, light even spells out a message. "Now D, now I, now L . . ." until this first neon sign reads, in Latin, "Love Justice, You Who Judge the Earth." A glowing eagle's head then appears, leading Dante to the Sphere of Saturn, where he sees a ladder, "its color, gold when gold is struck by sunlight." Flames descend the ladder in "so many splendors that I thought that every light / shining in the heavens was pouring down." A hundred suns then lead the poet into the Sphere of the Fixed Stars, where he sees "a sun above a thousand lamps." Questioned by St. Peter about his faith, by St. James about hope, by St. John about love, Dante is soon blinded by heaven's radiance. His vision is restored by light from Beatrice's eyes, "which shone more than a thousand miles." From there, Dante rises to the Sphere of the Angels, which Aquinas had called the Primum Mobile.

> I saw a point that flashed a beam of light
> So sharp the eye on which it burns
> Must close against its piercing brightness

As Dante trembles, nine blazing rings expand and resound with a hosanna. Angelic light bursts around him as he makes one final ascent, to the Empyrean. Here is Aquinas's "higher heaven . . . wholly luminous," and here the circles fade, leaving only a shining point and the glowing Beatrice. Encircled by radiance, Dante sees

> light that flowed as flows a river,
> Pouring its golden splendor between two banks
> painted with the wondrous colors of spring.
> From that torrent issued living sparks . . .

The sparks, he learns, are angels. Gazing still higher, he sees a lustrous rose, like that which stunned the first visitors to Abbot Suger's Saint-Denis. Beatrice leads Dante into the prismatic flower, where the pantheon of Christianity—the

Virgin Mary, Adam, Moses, and others—beckon him further. Finally, approaching the close of his epic journey, Dante sees what Moses was denied, what cathedrals only simulated: the essence of God. Unlike Prince Arjuna in the *Bhagavad Gita*, Dante does not tremble and look askance.

The living ray that I endured was so
acute that I believe I should have gone
astray had my eyes turned away from it.

At the apex of heaven, no earth-bound argument about light—Eye vs. Object, body or quality, infinite or measured—seemed to matter. Paradise, Dante assured the medieval world, possesses a light all-encompassing, all-powerful, all-knowing. Here is the light that shapes souls and drives the divine clockwork. And here the journey ends:

Here my exalted vision lost its power.
But now my will and my desire, like wheels revolving
With an even motion, were turning with
The Love that moves the sun and all the other stars.

By the time he finished *Paradiso*, in 1321, Dante's *Inferno* and *Purgatorio* had made him famous throughout Italy. While working on his masterpiece, he had moved to Ravenna, city of glittering mosaics, a fitting home for his final vision. A few months after he completed *Paradiso*, en route to Venice on a muggy summer day, he fell ill. Dante died that fall. *Paradiso* was published the following year. Within a generation, it spread throughout Europe, often in manuscripts copied by admirers. With its rhyming triplets and lilting language, each canto was what its name meant in Italian—a song. And the songs were soon memorized and read aloud in annual festivals. By the end of the 1300s, Dante's divine light—more portable than the Gothic, more universal than any architect's vision—rivaled that of the Bible itself.

In Canto XXIII of *Paradiso*, Dante claimed to be "figurando il Paradiso." Varying translations render "figurando il Paradiso" as "representing paradise," "picturing forth paradise," "describing paradise," and one, all too literally, "figuring of paradise." Yet Dante did not merely represent, describe, or picture paradise. He crafted heaven like an architect, shaping its ethereal rays into a light that endures.

The great Gothic cathedrals of Northern Europe now host nearly 40 million visitors a year, more than the combined attendance at the top ten

U.S. National Parks. Dante's trilogy sells strongly in dozens of editions, including audio books, e-books, graphic novels, video games, and iPad apps. "People can't seem to let go of the *Divine Comedy,*" the *New Yorker* noted in reviewing two new translations. The enduring appeal of divine radiance is measured not in lumens but in hope. Paradise and its luster urge the faithful to rise, to leave behind the mud and stones and doubt. The Middle Ages put this light on its highest pedestal. Each cathedral, each canto focused beams that still speak to those in need. Here is the realm that awaits. Here is its light.

CHAPTER 6

Chiaro e Scuro: *Light and Dark on Canvas*

Light is the first of painters.

—RALPH WALDO EMERSON

Every painter knows the importance of light. Just as vision depends on light, artistic vision demands its mastery. Until the invention of photography in the nineteenth century, only painting, as Leonardo da Vinci noted, could "portray faithfully all the visible works of nature." Yet faithful portrayal demanded a careful rendering of light, its subtle shadows, its infinite colors. Even in today's photo-drunk era, light taunts all who take up a brush. As early as elementary school, art teachers talk of perspective, shadow, and "light sources." Some students see these intuitively and win accolades. The rest struggle, shrug, and count the minutes until recess. Yet the vision needed to capture light on canvas took centuries, and perhaps a few tricks, to learn.

For well over a millennium, the light of daily life eluded artists. Greek lore tells of the artist Apelles, who painted grapes so convincing that birds pecked at them, yet no ancient Greek painting has survived. The walls of Egyptian tombs abound in magical figures rife with symbolic meaning, but no one pretends the world ever appeared so stilted, so unreal. By the Byzantine era, mosaics were catching light's sparkle, yet stones, no matter how small, could not pixelate the world with the precision of light. The most masterful mosaics depict folds in garments, the sheen of water, and minimal shading of the human body, yet their flat shadows render scenes as if from a child's pop-up book. Light was never this simple.

By the late Middle Ages, artists were again honing their talents with paint, yet they preferred *lumen*, the Latin term for radiance, to *lux*, the source of light itself. Whether reflected in gold leaf or rendered in a fresco's wet plaster, the light of medieval painting is filtered, its shadows scarce. Before light could be painted with subtlety, its own subtleties had to be noticed, yet not until the Renaissance did artists begin to appreciate how delicately "the first of painters" renders the world.

Throughout the 1400s and on into the next century, as the Renaissance spread from Italy north to the Netherlands, a handful of artists gave light a new incarnation—on canvas. Their expertise included all the fields a Renaissance man was expected to master, not just painting but geometry, chemistry, philosophy, architecture, and optics. Armed with the evolving tool of linear perspective and blessed with vision as sensitive as a camera's, these artists made photo-realist light the singular ingredient of Western art. These first masters of light included an embattled Dutch recluse, a murderer from the fetid alleys of Rome, and the greatest genius who ever lived.

Leonardo da Vinci did not discover how to imbue light's graces on canvas; he merely codified the steps. By the time of his birth, in 1452, light's artistic evolution had been under way for more than a century. Beginning in the early 1300s, the Italian master Giotto combined the beginnings of perspective with the first realistic gestures and backlighting, becoming, as the art historian Giorgio Vasari noted, "such an excellent imitator of Nature that he completely banished that crude Greek style." Giotto's lifelike frescoes were followed by the shaded biblical scenes of Masaccio and the spotlit *Annunciation* of Fra Angelico. Then, in the early 1400s, artists began using linear perspective. The technique of aligning the viewer's eye with receding "vanishing points" was perfected by Filippo Brunelleschi, who would later design the lofty dome of Florence's amazing cathedral. Painting by painting, master by master, Renaissance artists learned how to look, and look deeply, at light. Shortly before Leonardo's birth, they received their first guidebook.

On Painting named only one artist—Giotto—yet the widely read book exalted the art as "indeed worthy of free minds and noble intellects." "Painting," wrote Leon Battista Alberti, "possesses a truly divine power." To harness that power, Alberti advised, "painters should first of all study carefully the lights and shades." Alberti himself was not a great painter, at least not when compared with those he inspired. Though he considered himself chiefly an engineer and architect, Alberti was a consummate Renaissance man, writing poetry, plays, and art history, and dabbling in math and cryptography. Supremely confident and athletic, Alberti could ride wild horses and, it was said, jump over a man from a standing start. Along with the few buildings he designed, Alberti is best remembered for what he called "this little work of mine on painting." *On Painting* blended the optics of Aristotle and Alhacen with the aesthetics of the Renaissance to demystify the craft of catching light.

Writing as "a painter speaking to painters," Alberti began with simple steps. "Let me tell you what I do when I am painting. First of all, on the surface on which I am going to paint, I draw a rectangle . . ." Alberti adopted Alhacen's theory that light fills a room with overlapping pyramids. One of these strikes the painter's eye. A vertical plane slicing through the pyramid frames the visual phenomena we now take for granted: that shadows form on the side opposite the light source; that flat objects have a uniform light but spheres are subtly shaded; that painted objects should be circumscribed with the thinnest of lines. After describing these, Alberti took up the enigma that had perplexed light's earliest students: color.

Later painters would study rainbowed color wheels, but Alberti's advice was more elementary. First, he cautioned, avoid gold. Enough of the gilded Christs and Virgins that make some rooms in Florence's Uffizi Gallery glitter like jewelry stores. Only a few objects—sunlight, halos, the hair of certain women—should be golden, Alberti advised, and these should be rendered in paint rather than gaudy gold leaf. Once their palettes were muted, painters could appreciate

> a kind of sympathy among colors, whereby their grace and beauty is increased when they are placed side by side. If red stands between blue and green, it somehow enhances their beauty as well as its own. White lends gaiety, not only when placed between gray and yellow, but almost to any color. But dark colors acquire a certain dignity when between light colors, and similarly light colors may be placed with good effect among dark.

Alberti then focused on black and white. Only white conveyed "the brightest gleams of the most polished surfaces," and only black depicted "the deepest shadows of the night." Because the two capture the extremes of light, "those painters who use white immoderately and black carelessly should be strongly condemned."

Confident that he had done a "favor" to painters, Alberti concluded by asking for a modest homage. Any painter who appreciated his advice should paint his portrait among historical scenes "and thereby proclaim to posterity that I was a student of this art, and that they are mindful of and grateful for this favor." It is not known whether anyone painted Alberti in tribute, but like Dante, he had written in Italian so that all who could read might take his advice. Among the many who did was Leonardo.

His notebooks, written in his intriguing backward script, include sepia-toned sketches of the world in motion. Muscles and bones. Gears and levers. The flow of water, the prance of a horse, inventions to grind lenses, span rivers, take flight . . . Buried among all this ingenuity, deep in Leonardo's *Codex Urbinas*, is a single sketch that marks light's transcendence from cutouts to living masterpieces. Until Leonardo took up the study of light in 1490, its students had only traced its path, flat and angular, as if rendered by a clever child with a straight-edge and empty afternoons. Ptolemy and Alhacen imitated Euclid's schematics. Roger Bacon drew light rays that resembled strings. Light surely traveled along such angles, but it did not *look* like that. Then with one drawing, Leonardo did not merely tell us how light traveled; he showed us.

The sketch in the *Codex Urbinas* depicts a sphere struck by multiple light sources. Like track lighting along an arc, seven beams cast seven shadows. Angled behind the sphere are the pyramids Alhacen and Alberti imagined, overlapping, shading each other in a geometry Euclid could have described but never drawn. The sphere itself, depending on the angle of each light source, is layered from white to gray to black. Leonardo's drawing, combining the grace of an artist with the precision of a geometer, shows that light is never as elementary as night or day, *lux* or *lumen*, black or white. Leonardo's sketch *is* light.

When he began studying optics, Leonardo, not yet forty, had finished fewer than a dozen paintings. From his early *Annunciation* to the later Madonnas, each captured a light far more convincing than anything in Giotto, Masaccio, or other early Renaissance masters, yet each of Leonardo's paintings still showed the world through a glass darkly. The sky above Leonardo's *The Baptism of Christ* was an eerie violet. The blue of the Virgin's robe in the *Annunciation* gave her face a ghostly hue. Then, in 1490, Leonardo discovered a castle library in Milan that contained the works of Bacon, Alberti, and a Polish monk named Witelo, whose *Perspectiva* had popularized Alhacen. Diving into the works, Leonardo came to call optics "the blood of physics." He became enamored of perspective, which he called "*chiaro e scuro.*" (The translation of the Italian—"light and dark"—would later be condensed into the artistic term for stark contrast—*chiaroscuro.*) Throughout the 1490s, Leonardo put light under the microscope of his own eye.

His notebooks showcase his discoveries. On one page, a man in profile is struck by a pyramid of rays falling at precise angles on his chin, nose, and bald pate. Here, Leonardo writes, is "the Proof and Reason Why among the Illuminated Parts Certain Portions Are in Higher Light than Others." Elsewhere, light casts a gray oval onto an inclined plane and an angular shadow on stairs. "A shadow is never seen as of uniform depth on the surface which intercepts it unless every portion of that surface is equidistant from the luminous body." Further on, a robe spills off a platform and onto the floor, its every fold rendered in photographic detail. "You will note in drawing how among shadows some are indistinguishable in gradation and form." Leonardo's notebooks also describe how to sketch storms, battles, the seasons, and the night.

Studying his own eyes in a mirror, Leonardo was the first to notice how pupils dilate in darkness and contract at the approach of a candle. He speculated on the color of the sky, whose blue "is not its own color but is caused by the moisture that has evaporated into minute and imperceptible atoms on which the solar rays fall." He compared light to "a stone flung into the water [that] becomes the center and source for many circles." And in 1495, the year he began *The Last Supper*, he poured his theories into his *Treatise on Painting.*

Leonardo's advice to painters was that of a kindly teacher. Passages begin "O painter," and "You imitator of nature, be careful to . . ." He found light to be both lovely and puzzling: "Look at the light and consider its beauty. Blink your eye and look at it again: what you see was not there at first, and what was there is no more. Who is it who makes it anew if the maker dies continually?" Leonardo saw light as none had before him. The sky, he noticed, was layered and should be "lighter the lower you depict it." Black garments cast human skin in stronger relief than white garments. Shadows contained worlds of mystery. "When you represent in your work shadows which you can only discern with difficulty . . . you must not make them sharp or definite lest your work have a wooden effect."

Leonardo's *Treatise on Painting* changed the images of the Renaissance. Gone were floodlit canvases, solar spotlights, and portraits painted as if at high noon. "O painter, have a court arranged with the walls tinted black . . . or covered with a linen awning when the sun is shining; or else paint a portrait towards the evening or when it is cloudy or misty; and this is perfect lighting." Gone, too, were the saints in reds and blues, the landscapes in grays and greens. To highlight certain colors, he suggested, juxtapose them, for "every color is more discernible when opposed by its contrast than by one like itself." As an example, Leonardo cited the rainbow.

Though he admired Brunelleschi's linear perspective, Leonardo saw that depth depended on more than lines converging toward a vanishing point. Painters should enhance perspective by softening colors in the distance, just as light is softened by atmosphere. Remote objects lose their clarity and should be painted indistinctly. But Leonardo's most important art lesson concerned a phenomenon Alberti had noticed. In soft light, Alberti wrote, "color also dissolves progressively like smoke." Leonardo, using the same simile, perfected the technique of *sfumato*, Italian for "softened" or "blended."

Before studying optics, Leonardo painted his figures framed against dark backgrounds. This freed him from the childlike outlines so common in mosaics and other medieval art. His later paintings, done after 1495, display all that he had learned about light. *The Last Supper* frames Jesus and Judas in backlit windows. Mona Lisa and other subjects likewise blend seamlessly into backdrops smoothed by *sfumato*. To achieve the effect, Leonardo finished each painting with a thin coat of varnish made from cypress and juniper oils and laced with dark particles. Leonardo's amber varnish smudged all outlines, allowing him to "paint so that a smoky contour can be seen, rather than contours and profiles that are sharp and hard." This was how light rendered the world—neither stark nor golden but soft, subtle, and vaporous.

Leonardo planned to publish his *Treatise on Painting*, yet as scholars have long lamented, he had a hard time finishing anything. "Tell me," he often wrote, "if anything was ever done." Though never completed, his treatise seems to have spread like smoke. Throughout the 1500s, it surfaced in hand-copied notes and fractional manuscripts. Renaissance art's PR man, Giorgio Vasari, offered high praise: "Whoever reads these notes of Leonardo will be amazed to find how well that divine spirit has reasoned of the arts, the muscles, the nerves and veins, with the greatest diligence in all things." And from the high Renaissance until the dawn of Impressionism, Leonardo's "brief rules regarding lights" were those that artists followed. Of course, being artists, some broke the rules.

His given name was Michelangelo Merisi, but because Italy already had a famous Michelangelo—Michelangelo Buonarroti—Merisi made his hometown his signature: Caravaggio. As a rebellious, violent teenager, Caravaggio was shipped off to bustling Milan, where in 1584 he enrolled in an art academy. There he read Vasari's life of Leonardo, saw his *Last Supper*, and perhaps read his *Treatise on Painting*. Caravaggio also studied a little-known artist who only took to writing after going blind. For Giovanni Paolo Lomazzo, light was "the image of the divine mind," and an enigma that "has so great force in pictures that (in my judgment) therein constitutes the whole grace thereof." Lomazzo did not advise painters to mute natural light or paint late in the day. Instead, he suggested placing a lamp directly above a model, to create light like "that which the sunbeams make when they slide along upon the sea at sun rising." The visionary Leonardo had sought a light as blessed as a balmy afternoon but the blind Lomazzo preferred the light that pierced the night. And by the time he left Milan for Rome, the volatile young Caravaggio was well acquainted with darkness.

He had not been in Rome long before the name Caravaggio began appearing among art patrons and on police records. He was known for strutting the streets, shaggy and goateed, his sword and rapier close at hand. Rumors said Caravaggio had spent a year in jail, possibly for slashing a prostitute. He was frequently arrested for brawling, a common crime in Rome at the close of the 1500s. The city was enjoying a building boom that spawned dozens of villas, churches, and palaces. Meanwhile the surrounding countryside was in turmoil, its farms untended, its landowners bankrupted by papal taxes levied to fund the mightiest construction project of all, St. Peter's. Into the streets of Rome poured beggars, thugs, and thieves. Caravaggio, by all descriptions short and homely, bawdy and brilliant, fit right in, making Rome his set piece and personal den of sin. When the mood struck him, he painted.

Art historians still debate the metaphor suggested by every work by Caravaggio. Can his paintings, with their coal-black shadows and blinding light, be compared to his reputation for crime? "No one who loves Caravaggio," wrote biographer Howard Hibbard, "believes that his use of chiaroscuro was merely a technical device." The biographer Peter Robb went further: "The light and darkness of the world, he was reminding people, was also the light and darkness of the mind." One thing all agree on is that no artist before him had infused light with such raw power. Leonardo had cautioned against "too much light . . . too much dark," but Caravaggio scorned caution. From his first surviving work, *Boy Peeling Fruit*, his audacity was on display. The boy, a street urchin Caravaggio often used as a model, is caught in descending beams that streak across his white shirt, leaving the background as black as pitch. After *Boy Peeling Fruit*, Caravaggio briefly tamed his light, but by 1600, he was painting in a darkened studio with black walls, the only light coming through a window while he stood at his easel in shadows, real and imagined.

In a career that spanned just twenty years, Caravaggio shocked the public with scenes of androgynous boys, compromised women, and violence previously witnessed only in alleys. The biblical Judith holding the head of Holofernes was a common artistic motif, but Caravaggio caught Judith in the act of beheading. Abraham had stopped short of slaying Isaac, but Caravaggio painted the shiny blade just inches from the boy's naked throat. Even when painting a minor mishap, as in *Boy Bitten by a Lizard*, Caravaggio proved himself a master of a light so bright it hurt. Beside the grimacing boy stands a vase and in its glass shines the next two centuries of Western painting.

Though Leonardo knew the effects of luster, the shimmering aura "always more powerful than light," he preferred the light that fell softly across clothing and faces. Yet Caravaggio displayed—some might say "showed off"—a luster that turned fruit into satin and vases into tinsel. In *Boy Bitten by a Lizard*, Caravaggio renders a vase whose every reflection, every surface droplet is highlighted as if by the sun itself. Never had light been painted with such devotion nor in such sublime detail. Abbot Suger had seen God in the "most luminous windows" of Saint-Denis. Caravaggio's light came from a less holy source.

When finished with a painting, Caravaggio took to the streets, drinking, whoring, fighting. Once back in the studio, his portraits, like that of Oscar Wilde's Dorian Gray, reflected his darkening soul. Even as his name appeared more often on police records, his mastery continued to astonish. Behold the shimmering reflection of Narcissus gazing into his pool, the gleaming armor of soldiers capturing Jesus, the radiant faces staring into the divine light summoning St. Matthew. Together, these masterpieces made "chiaroscuro" a fixture of Western imagery.

By 1605, when his possessions were confiscated as payment for overdue rent, Caravaggio's fame extended to northern Europe. "A certain Michael Angelo van Caravaggio is doing wondrous things at Rome," an Amsterdam art critic wrote. "Already he has achieved with his works great repute, honor and a name . . . so setting an example for our young artists to follow." Several Dutch painters in Rome, known as the *Caravaggisti*, were already competing to capture that light. Soon another report from Rome clouded the aura surrounding their idol. "The painter Caravaggio has left Rome badly wounded, having on Sunday evening killed another person who had provoked him in a fight."

The fight was over a gambling debt. Surrounded by thugs, Caravaggio flashed his sword and cut a man down. Another sword came in high. Nursing a gash in his head, he fled the city. He spent the next four years in exile, still painting while dodging the law, including a papal price on his head. He died, alone, on a Tuscan beach in 1610, of malarial fever. His work remained popular in Rome for another decade, then fell into disrepute. By then, however, Dutch *Caravaggisti* had taken his style home, home to where light was being rendered with a precision that still stirs both awe and suspicion.

In February 2000, on a wall in his studio in Los Angeles, the British artist David Hockney began pinning up copies of paintings by European masters: Giotto. Masaccio. Van Eyck. Leonardo. Caravaggio. Rubens. Vermeer. When he was finished, Hockney's "Great Wall" stretched to seventy feet and spanned the last five hundred years of art history. Pacing before the progression, Hockney came to a controversial conclusion: something happened to Western art in the fifteenth century, something Hockney called "the optical look."

From the 1430s onward, Hockney asserted, Dutch artists used light to copy itself. Ever since the Mohists in China, all cultures had observed how light beaming through a hole projects a perfect image, inverted, on a wall. Alhacen had invented the camera obscura, whose wonders were widely known in fifteenth-century Holland, where lens grinders had made eyeglasses common and would eventually invent the telescope and the microscope. Surely, David Hockney suggested, Dutch artists must have used lenses to project light on canvas in order to paint its intricacies. To investigate further, Hockney studied the history of optics and of Western art. He noticed how, within a single generation, the smudged fabrics in a Masaccio evolved into the filigreed garments in a Van Eyck. He wondered how artists so quickly learned to foreshorten lutes and violins, or blur floral patterns in ways our self-focusing eyes could never notice. Weren't most artists known to possess mirrors and lenses, even depicting some in their paintings? And why did "the optical look" take hold only in Europe? Chinese or Indian artists should surely have evolved a similar vision—unless of course, their lenses could not compare to Europe's.

Hockney consulted with art historians, optics experts, and fellow artists. Some were curious, others shocked. How dare David Hockney, though an eminent artist in his own right, suggest that the likes of Vermeer and Caravaggio "cheated?" Hockney defended his thesis in articles and slide shows before museum crowds. In 2006, he made his case in a profusely illustrated book, *Secret Knowledge: Rediscovering the Lost Technique of the Old Masters*. Like a defender of Darwin, Hockney sequenced images to show the evolutionary leaps artists took between early Renaissance and late Baroque. The light of shining armor, barely a glint in fifteenth-century paintings, was captured a century later as if with a digital camera. An early lace collar was gray and fuzzy, yet a similar collar painted by Franz Hals, made without a single study or sketch, is starkly lit, perfect. And Caravaggio's chiaroscuro? He must have painted the spotlight and shadow captured by his camera obscura.

Hockney's thesis was not new. Art historians had long suspected certain artists of using optical aids. Canaletto's postcard landscapes of Venice made him a prime suspect. Another was Johannes Vermeer, whose portraits stream with light spreading from windows like lace. Some accused Vermeer of turning his studio into a huge camera obscura, letting the full landscape shine onto a canvas to paint *View of Delft*. Others noted how many of Vermeer's scenes are set in the same room with the same lighting. In 2000, the Oxford professor Philip Steadman created a mock-up of this room, reproducing all the illusions of an actual Vermeer. "The camera obscura served Vermeer as a composition machine," Steadman wrote. "Vermeer would have studied the camera image, the shapes of the objects and their shadows, the negative spaces between objects; and worked on the compositions by moving the objects themselves . . . His compositions are very far from casual snapshots." Other scholars aren't so sure, arguing that Vermeer was "a master of illusion, including the illusion that he had rendered scenes and optical incidents exactly as he had perceived them."

Leonardo had toyed with a camera obscura, yet stopped short of using one. "Such an invention is to be condemned in those who do not know how to portray things without it." The Leonardo scholar Martin Kemp, though assisting David Hockney's research, also remained skeptical. Written evidence of artists using the camera obscura "is almost entirely lacking," Kemp wrote. Visual evidence, while curious, "tends to be inconclusive." As the debate continued, Hockney insisted he was not accusing anyone of cheating. His suggestion that artists used optical devices "is not to diminish their achievements. For me, it makes them all the more astounding." As a lifelong lover of images and their creation, Hockney was "only saying that artists once knew how to use a tool, and that this knowledge was lost."

In their determination to capture every last highlight and glint, did some painters use light's own optical stunt? We will probably never be certain, but the controversy blurs the distinction between art and imitation. Camera obscuras, even under a spotlight, do not highlight fruit with the loving care of Caravaggio. Lenses can project an image, but they do not match precise colors. Between the dawn of Van Eyck and the heyday of Vermeer, the evolution of Hockney's "optical look" demanded more than devices. An artist might project light onto canvas and trace its intricacies down to the last fleur-de-lis, but capturing the details, the power, the soul of light demands the eyes of a master.

No one has accused Rembrandt of using a camera obscura, but once the world stood back from his paintings to fully appreciate them, other suspicions surfaced. That light, that golden, ethereal, shimmering light—surely there was some trick behind it. No eyes could be so sensitive, no brush so skilled as to create such a living illusion. The speculation began shortly after Rembrandt's death in 1669 and has continued into the twenty-first century. Rembrandt must have used pure gold in his paint. He must have employed some special binder. His varnishes must have aged, darkening the shadows, enhancing the radiance. His pigments, perhaps . . .

Since the nineteen sixties, scholars have used light to study Rembrandt's works. The Dutch-based Rembrandt Research Project X-rayed some 250 paintings. The X-rays showed only that Rembrandt worked from back to front, starting in the shadows, painting his luminous figures last. Infrared light, sensitive to heat, detected the charcoal beneath the pigment. Again, nothing suspicious. Gas chromatography was used to analyze Rembrandt's pigments, emitting telltale spectra revealing their chemical composition. That special binder Rembrandt must have used? It was linseed oil, sometimes thickened by egg yolk. His paint contained no gold. Samples of his varnishes revealed no unusual darkening agents. Perhaps the trick was in the artist's eyes.

They stare at us from his many self-portraits. Laughing in his youth, later joyous with his wife, Saskia, in his lap, finally becoming aged and remorseful, the eyes reveal little more than an artist being himself. More revealing than Rembrandt's own eyes are those in his other portraits. Close-ups reveal not just the sparkling pupils all artists pinpoint but subtle beiges beneath eyelids and flecks of moisture in the eye itself. To have seen so acutely, to know light so innately, surely suggests some secret.

A visit to Rembrandt's studio reveals how the Dutch master manipulated natural light, sometimes using a canvas in the corner to cover windows.
Photo by the author, used by permission of Rembrandhuis

The search for a secret to Rembrandt stems from the shroud he cast over his life. He wrote no artist's manifesto. His religion remains a subject of debate. He never traveled more than sixty miles from his birthplace and left only a few letters. His techniques, though extensively analyzed, have never been duplicated. Yet his skill was evident at an early age, convincing his father, a miller, to let him leave the University of Leiden for art school in Amsterdam. There one of Rembrandt's teachers, recently returned from Rome, taught him the art of chiaroscuro. Rembrandt's earliest paintings feature the black backgrounds and floodlit figures of the *Caravaggisti*. He would gather the rest of his light on his own. Opening a studio back in Leiden, the twenty-year-old painter did portraits and biblical scenes that possessed no special radiance. Only in 1629, when Rembrandt painted himself in his shadowed studio, did his fascination with light emerge.

Over the next decade, fascination grew into obsession. *Judas Repentant, Returning the Pieces of Silver* (1629) debuts the golden flash on a breastplate that would soon become a trademark. *St. Peter in Prison* (1631) adorns the aged saint in an earthly light that mocks any suggestion of the divine. And numerous portraits, their warm faces lit from one side, in a style now called "Rembrandt lighting," earned Rembrandt a steady living and a growing reputation. One of many who noticed was Constantin Huygens, a wealthy diplomat whose son, Christiaan, would later pioneer the wave theory of light. The elder Huygens, friend also to Galileo and Descartes, secured commissions for Rembrandt and bought several works. By then, Rembrandt was living in Amsterdam, at the home that still bears his name. In search of secrets, I paid a visit.

Some two hundred thousand visitors a year come to the Rembrandthuis. The elegant home overlooking a canal houses only one actual Rembrandt, yet its rooms seem like living canvases. Soft light suffuses kitchen, parlor, and especially the upstairs studio. Here, before a small crowd, the painter Eric Armitage explains the creation of a Rembrandt. Armitage begins by mixing paint, a chemistry lesson artists performed daily before the invention of tube paint in the 1840s. A spoonful of powdered pigment. A soupçon of linseed oil. Stir. Pour the goop onto stone and mash with another stone. Scrape and mash, scrape and mash. Finally, Armitage scoops the paint onto a putty knife and wipes it into a pig bladder.

Next came the preparation of canvas—or a wooden panel—which Rembrandt sealed with a chemical base. As the base dried, he would adjust the lighting. Because his studio windows face north, Armitage explains, the sun loops behind the house, barely changing the ambient light as the day progressed. To capture the right mood, Rembrandt might have merely closed his shutters, yet Armitage reveals another technique—a swatch of canvas pinned to the ceiling above a corner window. Here models posed beneath the light that Rembrandt could micromanage to perfection. And once that light was set, the eyes went to work. As X-rays have revealed, Rembrandt started with the recesses, spreading brown and ochre pigments up to a quarter-inch thick. Squeezing colors from pig bladders, adding shades and highlights, he strove for a light that, when seen from a distance, glowed like the world itself when seen from a distance. "My lord," he wrote to Constantin Huygens, "hang this piece in a strong light and where one can stand at a distance, so it will sparkle at its best."

The historian Simon Schama noted how Rembrandt defied Leonardo and Alberti. These Renaissance masters had suggested that softer colors make objects seem more distant, but "in a stroke of daring, Rembrandt has set the most brilliant areas at the rear, so that Delilah (in *Samson Betrayed by*

Delilah) seems to be fleeing the carnage out of the dazzling back of the painting." Rembrandt, one critic wrote, was "his own sun-god," and never was that god more powerful than in the painting housed about a mile—six canals—from his home.

Here in the celebrated Rijksmuseum, sound, not light, draws me toward Rembrandt's most famous work. Chatter and hum fill surrounding rooms and echo down the corridor that leads to *The Night Watch*. Hearing the noise, then spotting the painting through the portico that frames it, visitors walk swiftly past other Rembrandts and a few Vermeers to enter the throng. Standing ten rows deep, tourists murmur and gawk. School groups shuffle their feet as teachers explain, in Dutch, German, or French, why *this* painting has drawn such a crowd. Fingers point. Heads shake.

Is *The Night Watch* famous just for being famous? The enormous painting of a Dutch militia in shadows and spotlight surely has a *Mona Lisa*–like celebrity, yet those who wait their turn, inching closer as the curious move on, are drawn by more than mere fame. The gilded finery of one figure, the golden little girl to his side, the myriad faces in varying penumbra, all combine to create a light as natural as the stances of the subjects. Further study leads to surprises. The little girl has the same face as Rembrandt's beloved Saskia. A dog lurks in the gloaming. And far at the back, peering over shoulders as if at the rear of this same crowd, is Rembrandt himself, in cloth cap. His is the deepest shadow. In the teeming gallery, admirers jostle and shove forward, cameras held high. These photographers could easily download a higher-resolution image, but that would not be their Rembrandt, their own piece of him. Standing before *The Night Watch*, the crowd bears witness to mastery. They learn little about Rembrandt, however, other than that for one glorious year, when he was a new father, a loving husband, and in full command of his talent, he mixed his pigments, focused his sorrowful eyes, and made one perfect portrait of light.

The Night Watch was done both at the height of Rembrandt's career and on its precipice. By 1642, he and Saskia had lost three newborns, and also endured the death of his mother and her sister. Then, within weeks after *The Night Watch* was finished, Saskia died of tuberculosis. Unlike Caravaggio, Rembrandt did not let death darken his paintings. Instead, he turned primarily to drawings and etchings, which though gray at first, evolved to capture the shimmer of trees in sunlight, the spherical glow of a candle, and pyramids of light pouring in windows. By the 1650s, his self-portraits, once so confident and proud, showed a brooding man. Problems with Saskia's will led to bankruptcy, forcing Rembrandt to paint more. Now his subjects—

saints and apostles, his lone surviving son, the elite of Amsterdam—glowed like the embers of his life.

Rembrandt's final works, far from tragic, feature bright flesh and radiant garments. To pay homage, I leave *The Night Watch* and walk to the nearest alcove where an equally miraculous work hangs almost unnoticed. *The Jewish Bride* depicts a demure woman standing beside her husband, who has one hand on her breast. The husband's sleeve, layered thick with yellows and golds, blazes with all the light of creation. This was the light that stopped Van Gogh in his tracks. After staring at *The Jewish Bride* for an hour, Vincent told a friend, "I should be happy to give ten years of my life if I could go on sitting here in front of this picture for a fortnight with only a crust of dry bread for food."

Three of Rembrandt's last four paintings were self-portraits. The eyes in each are those of a survivor. Light was Rembrandt's Holy Grail, one he pursued for more than four decades. The eyes in his final self-portrait, weary yet proud, reveal the human toll of chasing such a chimera. They also suggest that no matter how lovely the radiance of this earth, no matter how boldly a Leonardo, a Caravaggio, even a Rembrandt might make it shine, light remains fleeting, ephemeral. And life itself is the trick.

CHAPTER 7

"Investigate with Me What Light Is": The Scientific Revolution and the Century of Celestial Light

Speak ye who best can tell, ye Sons of light.

—JOHN MILTON, *PARADISE LOST*

On an amber spring afternoon in 1629, high above the ancient ruins of Rome, five suns commanded the sky. The sun itself blazed in its usual place, but on either side, twin lights shimmered with rainbow colors. Above this trinity, like jewels in a tiara, two softer suns shone. Seen from the hillside of the Villa Borghese, the solar lights crowned the new dome of St. Peter's. Viewed from the Roman Forum, the five suns capped ancient porticoes and pillars. At the Vatican, the sight caused some priests to drop to their knees, others to shudder at this omen of five bad years. Merchants came out from their shops to gaze. Children pointed and wondered. The suns blazed for an hour, then faded until there was only one. Nine months later, on an icy January day, another crown of light appeared over the Eternal City. This one had seven suns.

Rome's solar spectaculars came during a remarkable astronomical era. This Century of Celestial Light started on a November night in 1572 when a new star appeared near the W shape of the constellation Cassiopeia. The era ended a hundred years later as European cities began hanging lanterns on dark streets. In between stretched an "age of the marvelous," a century filled with comets, supernovae, and other "starry messengers." It was a century of religious wars, but also of a peace that realigned Europe. It was the century when alchemy competed with chemistry, when astronomers unlocked the heavens but also cast horoscopes, when blood was first known to circulate yet the sick were bled until they swooned. It was a century of discovery, yet also of the Inquisition, witch hunts, and ignorance under papal decree. And it was the century that gave birth to modern science, using light as its midwife.

Light's departure each evening still spread terror. Dusk brought "the shutting in," the hour when walled cities closed their gates, shutters slammed, and all hurried home. Down black, cobblestone streets, thieves, madmen, and murderers roamed. "The night has fallen," a monk lamented, "covering the world with horrid darkness." An Italian proverb cautioned, "Who goes out at night goes looking for a beating." Each sunrise revealed the night's grim toll. Another body bludgeoned by the roadside, another bloated corpse in a river. Survivors of the night gave thanks to Light, the eternal blessing. But with the sense of inquiry ascendant, light soon transcended blessing, becoming the wonder of the age.

The "new star" of 1572 was not all that new. What we now call supernovae—exploding stars—were recorded by the Chinese as early as 185 CE and by Europeans every few centuries after. Yet this singular star, shining for more than a year, visible even at noon, was followed five years later by the most startling comet ever seen. Brighter than the moon, with a tail that stretched across a third of the sky, the Great Comet of 1577 led thousands to gather on hillsides, trembling, praying. What threat would the heavens pose next? For two decades, more comets flared and faded, punctuated by the usual eclipses and meteor showers like sparklers in the night. Then, in 1604, another new star burst forth. Three years later came another comet, now known as Halley's, then in 1618, three more. Finally came a spring day in 1629 when five suns shone above Rome. These celestial lights shook the assurance of the wise. Could the majestic clockwork still be trusted? Might there be new explanations for the universe? For light itself?

In the decades that followed the new star and the Great Comet, Renaissance men struggled to explain light, yet they were hampered by superstition, folklore, and the blend of both that was commonly called "natural magick." Two legends predominated. The first was of the Pharos light that once beamed over Egypt's northern coast. Earthquakes had since destroyed Alexandria's towering lighthouse, yet its ruins could still be seen in the harbor. This Wonder of the Ancient World had grown more wondrous in the retelling. Renaissance travelogues claimed the Pharos light had been visible for one hundred miles, even five hundred. And the mirror spreading such light had performed reciprocal miracles, so it was said. According to an Arab geographer, "When someone looked in the mirror he could see everything that was taking place in Constantinople, despite the fact that between this city and Alexandria there is the Mediterranean Sea and three hundred leagues." The Pharos lighthouse's mirror was rumored to be still more powerful. A 1550 travelogue claimed that "all ships passing near the column when the mirror was uncovered were miraculously and instantly burnt up."

This claim echoed the other enduring legend about light's magic, that of Archimedes. Immortalized in Greek and Roman lore, the story of his solar ray was told and retold, renewing interest in "burning mirrors." In the 1,700 years since Archimedes, many had speculated about such mirrors. As a young art student, Leonardo had seen "mirrors of fire" used to weld bronze sculptures. His notebooks featured burning mirrors beyond any yet known—huge parabolas twenty feet across. Scholars doubt that Leonardo ever made such a mirror, but late in the Renaissance, one man did.

A gentleman of Naples, Giambattista Della Porta wrote plays, toyed with inventions, and started his own Academy of Secrets. Brilliant, roguish, and romantic, Della Porta had a flair for collecting others' secrets and devising his own. His bulbous eyes scanned the world's wonders while his bald pate might itself have focused the sun's rays. Della Porta was best known for his twenty-volume series, *Natural Magick*, which sold out many printings. A lyrical synopsis of "all the Riches and Delights of the Natural Sciences," *Natural Magick* included chapters on "Changing Metals . . . Counterfeiting Gold . . . Tempering Steel . . . Beautifying Women . . ." The miscellany also featured tips on astrology, alchemy, perfuming, and how to make gunpowder and aphrodisiacs. One section described making a candle out of rabbit fat, a candle whose flame "constrains women to cast off their clothes and voluntarily shew themselves naked unto men." Just before light returned to the laboratory, such were the fantasies it inspired.

Beginning in 1580, Della Porta and fellow adventurers gathered at Venice's shipbuilding marvel, the Arsenal, to build a giant parabolic mirror. Such a mirror, Della Porta predicted, would outdo Archimedes, burning not "for ten, twenty, a hundred, or a thousand paces, or to a set distance, but at infinite distance." The mirror, however, proved disappointing. Unable to ignite ships, it was, Della Porta lamented, simply "an instrument for seeing far." Yet the gentleman from Naples continued to dream about light, imagining a mirror to project messages onto the moon, where they could be read by "a friend a thousand miles distant." And in 1589, his newest edition of *Natural Magick* added a book of optics.

Though his parabolic mirror may not have seen far, Della Porta's optics did. Twenty years before Galileo turned his telescope toward the moon, Della Porta explained how to make such a scope. The key was using two lenses, one concave, one convex. "With a Concave you shall see small things afar off, very clearly; with a Convex, things nearer to be greater, but more obscurely: if you know how to fit them both together, you shall see both things afar off, and things near hand, both greater and clearly." Della Porta never built a telescope, yet one of his plays features a character who buys one to spy on enemy troops. Taken to a rooftop, the man lifts the glass to his eye, only to see his daughter in her boyfriend's bedroom.

Della Porta's lofty prose captured the dream of Robert Grosseteste and Roger Bacon: of lenses that opened "the wonders of refracted vision." Within two decades of Della Porta's prediction, concave and convex lenses would beam light from distances no one could yet fathom. The telescope would make light the catalyst for an era that echoed Shakespeare's cry, "O brave new world!" But first, to amaze the trembling old world, the sky put on another show.

On an October night in 1604, high above the fairy-tale turrets of Prague and the bell towers of northern Italy, another exploding star blazed. Like the supernova a generation earlier, this one skirted the same W shape of Cassieopeia and was visible day and night. Galileo saw the new star from Padua, where he was teaching at the university. He calculated the star's parallax—its apparent shift as viewed from distinct points in the earth's orbit. Finding none, he judged this latest wonder to be far beyond the moon. But it was another astronomer—Johannes Kepler—whose name would be forever attached to the New Star of 1604.

Kepler was an eccentric for the ages. Awkward and shy, with pinpoint eyes and a Van Dyke beard, he was a blend of scientist, philosopher, and cleric. (He studied to be a Lutheran minister but was never ordained.) A mathematical wizard who proved that planetary orbits are elliptical, Kepler also made a good living by charting astrological calendars to predict weather and wars. Kepler considered astrology the mere "step-daughter of astronomy," yet he was sure the planets held some indefinable power over people. They had certainly messed with his family. The Keplers were a chaotic brood of fallen nobles whose manic intensity left a legacy of beggary, drunkenness, and witchcraft. Kepler's cantankerous mother, tried as a witch, narrowly escaped the fires of night; her aunt did not. As a child, Kepler was plagued by boils, sores, and "chronic putrid wounds in my feet," yet amid the chaos of his youth was a memory of being five years old and standing on a hilltop, gazing at a comet whose tail streaked across a third of the sky.

A few months before he saw the new star of 1604, Kepler published his first treatise on optics. The work was cluttered with the same "magick" that dogged Giambattista Della Porta. Kepler explained why comets spread illness, how planets affect the weather, and why storks raise their necks. But like the theologian he almost became, Kepler lauded light as "something akin to the soul," and God's embodiment on earth. "For when the most wise founder strove to make everything as good, as well adorned, and as excellent as possible, He found nothing better or more well adorned [than light], nothing more excellent than Himself. For that reason when He took the corporeal world under consideration, He settled on a form for it as like as possible to Himself." Despite his faith, however, Kepler did not endow light with divinity. Light was "the most excellent thing in the whole corporeal world," he wrote, yet in what should be read with a trumpet's blare, he announced that it "has passed over into the same laws by which the world was to be furnished."

Kepler harvested the fruits of Euclid and Ptolemy, al-Kindi, and Alhacen. His treatise on optics calculated eclipses, parallax, reflection and refraction, and the speed of light, which he judged to be instantaneous. Though wrong about light speed, Kepler was right about almost everything else. He updated Greek studies of mirrors, explored the camera obscura, and discovered a fundamental law of light. Why is light so blinding up close yet so weak across even a small room? Such a quality scarcely befits a "form . . . as like as possible" to God. Here Kepler built the bridge between the medieval and the modern. Setting aside theology, he drew rays at 45-degree angles emerging from the eye—the cone of vision. He then imagined a vertical plane slicing through the cone, creating a circle. Next, he applied the formula for the area of a circle, $A = \pi r^2$. The formula showed how a circle's area, A, expands exponentially as its radius, r, lengthens. If ordinary circles follow this formula, Kepler reasoned, a circle of light must spread and thin out at the same rate. When 2 feet from a candle, light forms a circle with a 2-foot radius. Its area is thus 4π, or about 12 square feet. But move the candle 6 feet away and the same amount of light forms a circle whose radius is 6 feet, with an area of 36π—113 square feet. Move the candle 10 feet away and the light is diluted across an area of 100π, becoming just one hundredth ($1/10^2$) as bright. This led Kepler to a crucial formula in physics: intensity = $1/\text{distance}^2$. Kepler had discovered the inverse-square law, a bedrock truth that Newton would apply to gravity and others would extend to all electromagnetic energy. The law explains why even a dozen candles don't fully light a room, why passing headlights look like pinpricks until they are upon you, and why all the lights of Manhattan make just a dim glow when seen from New Jersey.

Kepler's bridge to the modern was cluttered with the baggage of the ancients. His treatise referenced Aristotle and astrology, and explained how air is "poured about the moon." Yet in this cranky, clumsy, boil-ridden man, light had found its most calculating student since Alhacen. Another generation would pass before optics diffused the mysticism that had eternally enveloped light. Halley's Comet would come and go, as would three comets in a single year. But these three, unlike all previous celestial shows, would be witnessed through light's first technological wonder, the telescope.

No one is certain who invented the telescope, but its debut fulfilled centuries of yearning. Stories of all-seeing scopes date to Julius Caesar, said to have used a magic mirror in 55 BCE to see across the English Channel to the coast of Britain. More than a millennium later, the mythical kingdom of Prester John, whose existence was a matter of faith in the centuries before Columbus, was thought to be guarded by an all-seeing scope. Chaucer's "Squire's Tale" and Edmund Spencer's *Faerie Queen* fantasized about scopes that saw far, and the Jesuits, according to a seventeenth-century screed, had "a looking glass of Astrology wherein . . . there is nothing so secret, nor anything propounded in the privy councils of other Monarkes, which may not be seen or discovered." Giambattista Della Porta's *Natural Magick* made the dream imminent, but gathering light from afar required lenses of the finest quality.

In the three hundred years since Venetian glassmakers fashioned the first eyeglasses, lenses had barely improved. German craftsmen replaced Venice's ground-glass spheres with sliced glass fixed in wire frames. Gutenberg's printing press made reading more common, increasing the demand for glasses, yet lenses too often splayed light into haloes, spots, even rainbows. Light's scattergun rays could not be focused without distortion through one lens, let alone two. Then, in the autumn of 1608, three craftsmen working independently in three cities of the Netherlands filed patent applications for the first telescope. Hans Lipperhey had the greatest claim. To prevent haloes in his scope, Lipperhey devised a simple solution: a pinhole aperture placed over the leading lens to block the scattered rays and thread only the essential light.

In September 1608, Lipperhey demonstrated his invention during a peace conference at The Hague. Taking Dutch and Spanish generals to a tower, he had each lift the glass to his eye. Mouths agape, the men saw a clock tower in Delft, seven miles southeast, and the spire of Leiden's cathedral, fourteen miles northeast. The Spanish general said he would never again feel safe on a battlefield, and Lipperhey descended from the tower with orders for several more scopes. Before he could make them, other Dutch lens grinders presented their models. Letters spread the excitement throughout Europe. By April 1609, a shop in Paris was selling telescopes, and they were soon peddled in several other cities. That summer, word of these scopes, each with a magnifying power of three, reached Galileo.

As a devout Catholic, Galileo equated light with God (and sometimes with wine, which he called "light held together by moisture"). As a physicist he considered light "the universal starting point of nature." But these definitions would come later. First he would see light that none even dared to dream. When a friend told Galileo of the first telescope, he was "seized with a desire for this beautiful thing." Unable to find one in Italy, he had to make his own. He was, you might say, well prepared for the task. While teaching optics at the University of Padua, he had studied Euclid and Alhacen, Della Porta and Kepler. In the summer of 1609, Galileo took lenses from ordinary eyeglasses, ground one until it was convex, then inserted them at opposite ends of a lead tube. Within a day he had a working model, and within a week he had a telescope with a power of eight. That August he took his device to Venice. Leading stuffy old senators past the pigeons of San Marco and up the stairs of the Campanile, he had them aim his scope above the canals. And there in the glass were the white domes of San Giustino in Padua, twenty-five miles to the west. And there were the walls and ramparts of Treviso, far to the north. And there the sails of ships out at sea. And there . . . And there . . . Galileo's salary was doubled and he was given lifetime tenure. He went home to make a still stronger telescope, achieving a magnification power of twenty by mid-autumn. Then on a moonlit night in late November,

> Having dismissed earthly things, I applied myself to explorations of the heavens. And first I looked at the Moon from so close that it was scarcely two terrestrial diameters distant. Next, with incredible delight I frequently observed the stars,

> fixed as well as wandering . . . [and] to the three in Orion's belt and the six in his sword that were observed long ago, I have added eighty others seen recently.

In the summer of 1610, when Galileo's book *Starry Messenger* announced his findings, starlight became the stuff of daily conversation. No longer did people need comets or supernovas to inspire awe. Now the Milky Way, which Galileo revealed to be spattered with stars, offered fresh enchantment. "For the galaxy is nothing else than a congeries of innumerable stars distributed in clusters," Galileo wrote. The moon had mountains like the earth, he told Europe, and Jupiter had its own moons, pinpoints that circle the planet, now on one side, now the other. Galileo thanked God "that He has been pleased to make me alone the first observer of amazing things which have been obscured since the beginning of time." And Europe thanked Galileo, hailing him as a new Columbus. Kepler waxed rhapsodic. "O telescope, instrument of much knowledge, more precious than any scepter!" Generals bought the latest models and royal families begged Galileo to name stars after them, as he had named Jupiter's moons after members of the Medici family.

Galileo's light made him famous, yet fame would exact its cost, leading to his infamous papal trial and renunciation of the Copernican system. "Eppur si muove," he allegedly said of the earth that the pope fixed at the center of the universe: "Still, it moves." And on it moved, this earth, this century, changing how light was seen and studied. Once Galileo sent his starry message, ancient assumptions about light began to flicker and fade. Aristotle's platitudes were the first to fall. Aristotle had seen the moon as a polished surface, a mirror reflecting the sun, but Galileo challenged this by hanging a mirror on a wall. He described the scene in a dialogue between two searchers he named Sagredo and Salviati. If the moon were a mirror, Sagredo concluded, "it would be of an absolutely intolerable brilliance." Nor would a mirror hung in the sky be visible everywhere on earth. A mirror's rays, after all, struck just one spot on a wall.

For the rest of his life, Galileo pondered light. He knew how it bounced and bent, but what was it? Perhaps when two hard bodies rubbed together, he speculated, "when their ultimate and highest resolution into truly indivisible atoms is reached, light is created." If light was atoms, he realized, it would have a definite speed. To clock it, Galileo described Sagredo and Salviati standing with lanterns on hilltops a mile apart. Sagredo flashed his light, Salviati saw it and waved his. Given that light would travel the mile and back in 0.000005 seconds, it was not surprising that Salviati wondered whether "the opposite light was instantaneous or not; but if not instantaneous it is extraordinarily rapid." Galileo never performed the experiment, leaving light's speed—if it had one—to speculation.

"What a sea we are gradually slipping into without knowing it!" Salviati exclaimed.

"Really," Sagredo replied, "these matters lie far beyond our grasp."

Galileo was ultimately defeated by light's mysteries. Blinded by glaucoma, under house arrest, he shared his frustration with a friend: "I had always felt so unable to understand what light is, that I would have gladly spent all my life in jail, fed with bread and water, if only I was assured that I would eventually attain that longed-for understanding." Galileo died in 1642. By then, the most dazzling show in this Century of Celestial Light had inspired the man who would send light shining into the future.

Long before five suns shone above Rome in 1629, observers had seen the sky play similar tricks. Aristotle had written of "mock suns" that "rose with the sun and followed it all through the day." Others had puzzled over what we now call "sun dogs." In 1461, during England's War of the Roses, three suns appeared before a battle, inspiring Yorkist troops to victory and summoning Shakespeare's homage in *Henry VI.*

> EDWARD: Dazzle mine eyes, or do I see three suns?
> RICHARD: Three glorious suns, each one a perfect sun
> Not separated with the racking clouds
> But severed in a pale clear-shining sky.

But in the 150 years since the apparition, even the most devout cleric sensed how science had matured. Before the crown of suns faded over the Vatican, Cardinal Francesco Barberini sketched them. He might have given his diagram to his uncle, the pope, yet the cardinal already knew what the church would say. Instead, he sent his drawing to a prominent French astronomer. The astronomer copied the cardinal's diagram, mailed copies to friends, and throughout that summer the learned of Europe puzzled over the several suns of Rome. Most merely talked, but René Descartes went to work.

When he heard about the five suns, Descartes was living in Amsterdam, about a mile from the house Rembrandt would soon purchase. At thirty-three, the smug, witty Frenchman had impressed intellectuals back in his native France. Already the premiere mathematician of his age, Descartes had begun to dabble in philosophy and astronomy. But ten years had passed since he renounced superstition and certitude, making "a firm and steadfast resolution . . . never to accept anything as true if I did not know clearly that it was so." Later, Descartes would refine this resolution into his *Discourse on the Method* whose famous assertion "I think, therefore I am," jumpstarted modern thinking by replacing blind acceptance with skepticism and proof. Yet by 1629, after joining the army, wandering across the Alps, exploring "the great book of the world," Descartes had published nothing. He was laboring over a treatise on metaphysics when he learned of Rome's five suns. Setting the work aside, he decided to study light.

Why light? Descartes explained: "Just as painters, not being able to represent all the different sides of a body equally well on a flat canvas, choose one of the main ones and set it facing the light . . . so, too, fearing that I could not put everything I had in mind in my discourse, I understood to expound fully only what I knew about light." Descartes expected to compose a short study but he spent the next four years writing *The World, or A Treatise on Light*.

Descartes began by urging his readers to "investigate with me what light is." Peering into a blazing log in his fireplace, he saw particles in "very violent motion." Since fire gave off light, its particles suggested the essence of light itself. "The same motion which is in the flame," Descartes concluded, "is enough to make us have a sensation of light." Striding past Kepler, whom he called "my first master in optics," Descartes then defined thirteen properties of light. Light "extends in all directions about bodies one calls 'luminous.'" It travels "in an instant . . . ordinarily in straight lines." Such lines could diffuse or be focused by lens or mirror, on a burning point. Descartes considered light to be pressure, one that could be bounced, bent, or blocked. Consider it like a tennis ball, he explained. Traveling through air, a ball—or beam—strikes an object that reflects it at a precise angle. Water or glass reflect some light but also let some pass, deflected as if the tennis ball broke through a flimsy sheet and traveled on, its direction shifted in the collision.

Pondering how light transmitted pressure, Descartes revived the ancient Greek concept of "aether." Aristotle, insisting that nature abhorred a vacuum, reasoned that *something* must fill the vastness of space as well as the space between eye and object. Descartes saw ether, which he called "plenum," as that something. Light was ether's pressure on the eye, sensed by the wonderworks of vision. (To explain vision, Descartes dissected a cow's eye, slicing off the retina to use the lens as a gooey camera obscura. Don't try this at home—I did. The dead eye inverts an image, but also turns your stomach.)

Explaining light in terms so simple that, as he told a friend, even women could understand, Descartes summoned the common terror of night. "It has sometimes doubtless happened to you, while walking in the night without a light through places which are a little difficult, that it became necessary to use a stick in order to guide yourself; and you may then have been able to notice that you felt, through the medium of this stick, the diverse objects placed around you." Light's transmission to the eye, he continued, was like the tremor of the stick in hand. As to how light traveled, Descartes shifted from tennis balls to wine in a vat of grapes, the liquid seeping around the fruit to flow through a hole in the bottom. "In the same way, all of the parts of the subtle material, which are touched by the side of the sun that faces us, tend in a straight line toward our eyes at the very instant that we open them."

Having defined light, Descartes dared to examine the rainbow. Myth and religion had long lent purpose to the rainbow. Babylonians saw it as the necklace of the goddess Ishtar. In Genesis, the rainbow sealed God's promise to Noah that "never again shall there be a flood to destroy the earth." Aborigines spoke of a Rainbow Serpent, a giant snake that sometimes rose into the sky. Homer also exalted Zeus's "wondrous bow." In *The Iliad,* after the rogue Paris steals the lovely Helen from Sparta, Hera sends the messenger Iris to fly along the rainbow and tell King Menelaus of the outrage. Aristotle, however, would stand for no such nonsense. In his "Meteorology," Aristotle diagrammed a man eyeing sun and clouds, suggesting the rainbow as "a reflection of sight to the sun." Alhacen disagreed, arguing that the rainbow was formed when sunlight struck the concave surface of a cloud. The confusion continued into the Middle Ages. Then, in 1235, Bishop Robert Grosseteste suggested the rainbow was made not by bouncing light but by bending it. Refraction, not reflection. Roger Bacon took another giant step, estimating the angle of the rainbow above the horizon—42 degrees. Finally, at the start of the 1300s, two men separated by continents and culture yet linked by reading Alhacen figured it out. The Dominican friar Theodoric of Freiberg and the Persian scientist Kamal al-Din al-Farisi each proposed that raindrops both refract and reflect. Sunlight bends when entering a raindrop, reflects at a right angle off its inside wall, then bends again on exit. And this is just how a rainbow is made: two refractions, one reflection, and a bow of colors—hardly less miraculous than a promise from God.

Descartes, having seen rainbows in fountains struck by sunlight, used a fish bowl as his raindrop. With his back to the sun, he held his bowl aloft. On the opposite wall he saw a shimmering spectrum, red to violet. Descartes measured the refraction and, as the premiere mathematician of his age, devised a formula. Fifteen hundred years earlier, Ptolemy had calculated refraction with ordinary arithmetic, but he was as wrong as he was about the sun revolving around the earth. Refraction is not derived by arithmetic but by trigonometry. Descartes figured out, as had a Dutch astronomer and an Arab mathematician, how to calculate the angles. The result was light's first unit of optical measurement, the refractive index.

The denser a substance, the more it refracts light. The measure of how much a substance—water or olive oil or diamonds—bends light is its refractive index. A higher refractive index means a more pronounced bend as light slows upon entering a thicker substance. Water's refractive index is 1.33, olive oil's is 1.47, diamond's 2.42. The index uses trigonometry's sine function, a decimal obtained by dividing a right angle's hypotenuse by the side opposite the angle. Say I shine my laser pointer at a pan of water, the beam striking the surface at a thin 15-degree angle from the perpendicular. (In other words, my pointer is nearly straight up and down.) I divide the sine of the thin angle above water by the sine of the thinner angle made by the red beam through the water. The answer will always be 1.33. Knowing the index, doing the math, I calculate that light striking water at a 15-degree angle will bend to form an 11-degree angle underwater. In naming this formula, scientists added Descartes's name to that of the Dutch astronomer, Willebrod Snell, subtracted, or neglected, the Arab discoverer, and codified the Snell-Descartes law of refraction.

With the sine table ready, Descartes measured the angle made by sunlight and his line of sight while gazing at the rainbow. Sunlight hitting the rain formed the angle's upper ray. Descartes's line of sight formed its bottom ray. The rainbow's peak was the vertex—where the rays met. When the angle thus formed was precisely forty-two degrees, just as Roger Bacon had estimated, the rainbow appeared. Descartes's measure explained why the rainbow is a morning or afternoon phenomenon. Noonday sun never strikes rain at forty-two degrees. His calculation also explained why every rainbow appears at the same height above the horizon and why there is no pot of gold, nor even a rainbow's end. No matter how fast you approach them, the colors recede, maintaining their perfect angle.

After calculating a single rainbow, Descartes tilted his bowl to measure a rarer spectacle—the double rainbow. At forty-two degrees, he saw one such rainbow. But ten degrees higher, he often saw a secondary spectrum fainter than the first. For reasons he never fathomed, this second rainbow is inverted, the red appearing below, not above the violet. Yet why should a raindrop make colors at all? Here Descartes had to guess. If light's particles were tennis balls, each would spin. Passing through a raindrop, some would speed up, others slow down. The slower would be cast upon sky as red, the faster as violet. Descartes was wrong, doubly wrong. His first error was in believing that light traveled faster in water than in air. Air, he posited, was soft and squishy, slowing light just as a carpet slowed a rolling ball. "The harder and firmer are the small particles of a transparent body the more easily do they allow the light to pass," Descartes wrote. His second error was in assuming that red "tennis balls" traveled slower than violet ones. Yet despite his mistakes, Descartes had transformed the rainbow from a mystery into a math problem. He was almost finished with light. What about those five suns?

Though schooled by Jesuits, Descartes refused to see the suns as divine omens, or even suns. Instead, blending optics with meteorology, Descartes realized what had occurred in the sky above Rome. A warm wind off the Mediterranean had met a colder one blowing from the north. The two winds formed a floating ring of ice, each crystal acting as a prism. Using his law of refraction, he calculated the exact angles that would form five suns in a celestial crown above St. Peter's.

In a little more than one hundred pages, Descartes's *The World* demystified light. "I do not believe it is necessary for me to converse further with you," he told his readers, "for I hope that those who have understood all that has been said in this treatise will, in future, see nothing in the clouds whose cause they cannot easily understand, nor anything which gives them any reason to marvel." Descartes was wrong again. Centuries later, the complex science of optics has yet to dampen the marvel of light.

Descartes never investigated light again, preferring the simpler task of proving the existence of God. By the time he died in 1650, the sky had settled down. Comets came and went, yet none compared to the Great Comet of 1577. And after two new stars within a single generation, no supernova was seen without a telescope until 1987. But Descartes had taken light down from the heavens and put it on a sine table. There would be further arguments, philosophical and mathematical. The British philosopher Thomas Hobbes agreed that light was pressure but saw it as created by a pulsing sun. The French mathematician Pierre Fermat challenged Descartes's notion that light accelerated in water. Fermat's "principle of least time" suggested a lifeguard heading for a drowning swimmer. To reach a given point, Fermat argued, light sprints along the sand (air), swims more slowly in the water, but always reaches its destination by the quickest possible combination of the two speeds. But Descartes had changed the conversation. The devout would still see light as ethereal, yet others would continue Descartes's investigation of "what light is." And as the investigation deepened, darkness surrendered some of its terror.

Throughout the Century of Celestial Light, night remained nasty, brutish, and far too long. On an average night, along the dark streets of Paris, fifteen people were murdered. Behind castle walls, however, royal engineers were learning to illuminate huge ballrooms, making night-time concerts occasional features of French court life. One such performance was the *Ballet de Nuit*. Staged in 1652, this *Ballet of Night* saw a young Louis XIV costumed as a frilly, golden sun. Forever after, Louis would be the Sun King. Enchanted by light, he began hosting regular *soirées* lit by thousands of candles. Then, in 1666, on advice from his council of police, the Sun King decreed that hundreds of lanterns be hoisted by pulley and suspended by ropes to illuminate the boulevards of Paris.

There had been earlier street lighting, notably in Arab cities around the time of Alhacen, but nothing like this. Visitors were astonished. "The most distant peoples should come and see the invention of lighting Paris during the night with an infinity of lights." City officials came and many returned home with plans. Street lanterns were installed in Amsterdam (1669), Hamburg (1673), Berlin (1682), Copenhagen (1683), London (1684), Vienna (1688), Hanover (1690), Dublin (1697), and Leipzig (1701). The result, in cities at least, was the conquest of the "darksome monster," and a revolution in human behavior. Hordes of men (men only) emerged into once forbidding streets. Taverns and pubs stayed open toward midnight, coffeehouses flourished, and the idle rich became idler still. A French abbot recalled: "Before this age everyone returned home early for fear of being murdered on the street, which redounded in favor of one's work. Now, one stays out at night and works no more."

Under calculating eyes, light had been examined and measured. At the command of kings, it had begun to tame the night. No more would any but the most mystical speak of "natural magick" or of candles "to make women cast off their clothes." The stage was set, and lit, for Isaac Newton.

CHAPTER 8

"In My Darken'd Chamber": Isaac Newton and Opticks

Even Light itself, which everything displays,
Shone undiscovered, till his brighter mind
Untwisted all the shining robe of day . . .

—JAMES THOMPSON, "TO THE MEMORY OF SIR ISAAC NEWTON"

Late in the summer of 1664, a foppish Cambridge University student with shoulder-length locks left his lonely dorm to visit a fair on the Stourbridge Common. With England languishing between a civil war and a plague, the Stourbridge Faire was more crowded than usual and the student walked unnoticed past women in straw hats and wigged men quaffing pints of ale. Since its inception during the Middle Ages, the annual fair had grown into Europe's largest, teeming with jugglers, minstrels, and merchants. Odors of cheese and roasting geese filled the air. Hundreds of wooden booths brimmed with textiles, pewter, pots and pans, spices, silks, and draperies, all competing for attention with the maypole at the center of the common. Isaac Newton, however, was looking for something more modest. At the previous year's fair, he had bought a book on astronomy but found its trigonometry daunting. This summer, having just read Descartes's treatise on light, he stopped at a booth selling toys to buy a small prism.

In the course of his studies, Newton would write some four million words, yet scholars still don't know whether he bought his prism before or after he stuck a needle in his eye. The needle, a curved sewing tool the British call a bodkin, fit easily between eye and socket. Newton described his experiment: "I took a bodkin and put it between my eye and the bone as near to the backside of my eye as I could & pressing my eye with the end of it (so as to make the curvature in my eye) there appeared several white, dark and coloured circles." With the bodkin jutting from his uplifted face, Newton tested his tool. If he wiggled the steel shaft, he saw more circles. If he held eye and bodkin still, the circles disappeared. After a few moments, Newton eased the shaft out of his eye socket. Looking at a dark wall, he still saw circles, "a motion of spirits." When these vanished, he sketched the bodkin as it skimmed his eye and skirted his optic nerve.

The experiment deepened Newton's suspicions about Descartes, whom he called "Des-cartes" or just "Cartes." The Frenchman was wrong about light. If it were like vibrations from a stick, "then we should see in the night as well or better than in the day." If light were a pressure, we would see it above us "because we are pressed downward." By simply moving against light's pressure, "a man going or running would see in the night." But if Descartes was wrong, what was light? Sometime before or after buying a prism, Newton decided to stare at the sun.

Popular stories surrounding Isaac Newton—the apple, his lifelong virginity, his standing "on the shoulders of giants"—scarcely describe his impact. If a solar crown like the one that appeared over Rome were to depict light's pantheon, Euclid and Alhacen would be the lower two "mock suns," Einstein and James Clerk Maxwell the upper two. Newton would be the sun. His exploration of light was not as intricate as his creation of the calculus, nor as flawless as his laws of motion, yet Newton brought optics into everyday life. Any fool at a fair could buy a prism—since the heyday of Rome, the colors streaming from triangular glass had delighted children and intrigued adults—but it took Newton to unlock light's crucial secret with a toy.

Newton once spoke of being a child at play on the seashore, the vast ocean of knowledge beyond his reach. Yet he never saw the sea, suffered a joyless childhood, and was hardly as modest as his metaphor suggested. Nothing in his early life augured more than common drudgery. When Newton was born, on Christmas Day in 1642, his father, an illiterate farmer, had already succumbed to fever. Newton's mother, remarrying when Isaac was three, handed him over to his grandmother. Isolated and angry, Newton sought solace in notebooks he began filling at an early age. One read: "A little fellow; My pooer help; He is paile; There is no room for me to sit; In the top of the house-In the bottom of hell; What imployment is he fit for? What is hee good for?" Throughout his childhood, England fought with itself, Royalists and Parliamentarians battling in a civil war that had boys at Newton's boarding school choosing sides. The little fellow preferred solitude and observation. He might catalogue his own sins—"Punching my sister . . . Falling out with servants, Beating Arthur Storer"—but he found colors more alluring. What caused red, green, or "blew"? Were they inherent properties or mere perceptions? Equally enthralled by alchemy, he mixed foul potions and drank them to ease the aches he would suffer all his life. At seventeen, lacking any distinction in school, he was brought home to tend his mother's cottage in Woolsthorpe, a week's carriage ride north of London. But he roamed the fields, thinking, dreaming, until he was sent to Cambridge. He entered Trinity College in 1661.

Like most young men of his time, Newton's command of mathematics did not go much beyond Euclid, but he was surely the quickest study in the history of numbers. At Cambridge, he absorbed the analytical geometry of Descartes, the optics of Kepler, the physics of Galileo. He read England's latest work on optics, *Experiments & Considerations Touching Colours*, by Robert Boyle. Sometime during his studies, alone in his room with just a few candles, a desk, and a chamber pot, he stuck a bodkin in his eye. His eye was unharmed but when he held a mirror to the sun and stared at it, he was blind for three days. For the next several months, he saw colors and fireballs of light at odd hours.

In the summer of 1664, Newton bought a prism at the fair and sometime that fall, in his Cambridge dorm, began streaming light through it. From Robert Boyle he had learned that colors appeared when white light is "modified" by reflection or refraction, so Newton set out to modify his own light:

> Having darken'd my chamber, and made a small hole in my window-shuts, to let in a convenient quantity of the Suns light, I placed my Prisme at its entrance, that it might be thereby refracted to the opposite wall. It was at first a very pleasing divertissement, to view the vivid and intense colours produced thereby, but after a while applying myself to consider them more circumspectly, I became surprised to see them in an *oblong* form . . .

Alhacen, Descartes, and others had used prisms shorter than Newton's, each forming a rainbowed circle. Puzzled by his oblong spectrum, Newton measured the colors on his wall, finding them five times longer than wide, "a disproportion so extravagant, that it excited me to a more then ordinary curiosity of examining." Suspecting a faulty prism, he found another, let light through his "window-shut," and there again were the colors, in neat rows. He measured the distance to the wall and the angle at which light struck the prism (63 degrees, 12 minutes). Using the law of sines he had learned from Descartes, he calculated the prism's refraction index, which he called its "refrangibility," then measured the index of several substances—air, rain water, gum arabic, "oil olive." But if light was indeed like the tennis balls Descartes described, how did it curve into colors? He suspected the "ambient Aether" might refract red light more than blue, yet "I could observe no such curvity in them."

Newton was soon distracted by comets. During the winter of 1664–65, one comet kept him up all night, watching, wondering. A second soon appeared, and months later, a third. That summer, the plague gripped London, killing thousands, then tens of thousands. Trinity College closed. The Stourbridge Faire was canceled, but Newton no longer needed its wares. By fall, he was back at his mother's cottage, alone. Throughout the next year, which many call his annus mirabilis, Newton ate only when forced, living mostly on bread and water. Holed up in "my darken'd chamber," he pondered first a new branch of mathematics he called fluxions, then a daring theory about light.

As early as 400 BCE the Greek atomist Democritus asked, "What is redness?" Does an apple *seem* red because the eye perceives the color, or is an apple made of red atoms? Plato considered color to be "flames which emanate from all bodies, having particles corresponding to the sense of sight." Aristotle thought otherwise. Because the eye does not emit light, Aristotle said, it does not emit color, whose perception depends solely on brightness. Aristotle's views held for a few centuries before being challenged by the Roman poet Lucretius. Atoms, Lucretius wrote, "have no need of color."

The atoms are not clothed in any
color, it is clear.
What color can there be in blinding
Darkness? It's the play
Of light itself that changes colors,
Depending on the way
Its brightness bounces off them,
At an angle or direct.

On through the first millennium CE the confusion continued. In his Cairo study, Alhacen held different fruits up to sunlight, studying their changing hues. Color, he decided, did not depend on the strength of light, nor on its play. It was internal. As example, he cited blushing. When a face blushes red, Alhacen wrote, such color "has no cause other than that shame. And shame is not something that comes from outside, nor is it related to the light or the eye that looks at that face." By the Middle Ages, scholars were studying contrasts and blends. Roger Bacon noted how a spinning top mixed its colors. Another clerical scholar, John Pecham, observed how colors "appear varied according to the lights shining upon them." A color "in sunlight is wholly different from that in candlelight," Pecham wrote in 1277. "Furthermore all colored things are deprived of the customary beauty of their color during a solar eclipse."

By the time Newton bought his prism, the enigma of color remained. While at Cambridge he speculated that color depended on "imagination and fantasy and invention." Now, in 1666, while the world turned beneath the sun's steady light, while London suffered its plague, this twenty-four-year-old loner made light his servant, his magician, his plaything. Still puzzled by the oblong spectrum, he rotated his prism, marveling as the livid rectangle rose and fell, casting red light ascending, blue on its downward drift. He caught the spectrum on paper, moving it nearer, farther, noting that when blue was "distinct," red was "confused," and vice versa. If he tilted the paper, the colors stretched and paled. "As curious as could be" he extended the prism's beam across the dark room. The spectrum spread. He changed the size of the hole in his window-shut, then tried a thicker prism, but could make "no sensible changes in the length of the image." Finally he began what he called the *experimentum crucis*. Any fool could stream light through one prism, but Isaac Newton used two.

He places his second prism across the room. A full spread of colors lights the wall. Lifting a wooden board drilled with a hole, he blocks all colors but red. A single scarlet thread strikes the second prism. He expects the glass will diffuse this red back into white but the shaft beams through, onto the wall. Red. Still red! Tilting the prism, he sees orange do the same. Light, he now knows, is made of all colors, and once separated, each is pure and distinct. Soon he is shifting his prisms, putting light on parade—red, yellow, green, blue, violet. He cuts a paper circle and hits it with blue from one prism, red from another, turning the paper a bright purple. Like Rembrandt he mixes his colors and like Alhacen, he records the results: "Red and blue make purple. Yellow and red make orange. Purple and red make scarlet. Red and green make a dark tawny orange." Like Galileo fashioning his own telescope, he is soon making prisms, gluing glass plates together to fashion triangular tanks that he fills with rainwater. And like Newton alone, he sees—"Hence redness, yellowness &c are made in bodys by stoping the slowly moved rays Without much hindering of the motion of the swifter rays, & blew, greene & purple by diminishing the motion of the swifter rays & not of the slower."

He will never be done with light, never explain it fully. He manipulates colors but will never learn how to handle people. Decades of bitter feuds will follow these revelations. Resentments will rise, float, and descend like the colors on his wall. Churlish and vindictive, he will refuse to publish his treatise, *Opticks*, until the next century. But ahead also lies the light whose mastery he enabled, light that, reframed by relativity and quantum theory, now reads and heals, measures and maps, but still behaves much as Isaac Newton said it would.

The first people to hear of Newton's revolutionary discovery about light could not have cared less. In January 1670, recently named Lucasian Professor of Mathematics at Trinity College, Newton began weekly lectures. His predecessor had finished the previous year speaking of optics, so Newton picked up the thread. Standing in the frigid classroom, its wooden tables creaking, its light limited to what came through the window on a winter day in England, Professor Newton began. "The recent invention of telescopes has so occupied most geometers that they seem to have left to others nothing in optics untouched nor any room for further discovery . . ." Perhaps there was a yawn, or just silence. The lecture was in Latin, after all. Newton continued. "It might perhaps seem a vain endeavor and futile effort for me to undertake to treat this science again. But since I observe that the geometers have hitherto erred with respect to a certain property of light pertaining to its refractions . . ." Newton lectured for an hour. He was not shy about claiming his discovery, telling students of his experiments "lest you think that I have set forth fables instead of the truth." A week later, when he returned to continue his class, the room was empty. He lectured anyway, as he often would throughout his academic career, reading, as a colleague noted, "to the walls." Students cared nothing about Newton's discovery. As with all of light's enigmas, the answers interested only the most curious characters.

Since 1660, members of the Royal Society of London for Improving Natural Knowledge had been meeting in a large, wood-paneled room at Gresham College, not far from the Tower of London. There, before a roaring fire, these pompous men in flowing wigs and frock coats behaved much like boys tinkering with nature. Each meeting began with studies—of tides, weather, the phases of the moon—but then came the experiments. The Royal Society was not in session long before its members examined the horn of a unicorn. The popular field of teratology—the study of deformities—was a special favorite, leading members to bring in two-headed calves, monstrous stillborns, and enormous tumors. Women said to have been pregnant for years were a frequent source of study, and heaven help the poor animals—dogs, snakes, chicks—who succumbed to the latest vivisection techniques right there before the open fire.

To improve natural knowledge, the Royal Society also published the world's first scientific journal. Its title, *Philosophical Transactions*, suggested that the distinction between science and philosophy still awaited Isaac Newton. And in 1671, when Newton sent the society a new marvel, a six-inch reflecting telescope more powerful than Galileo's refracting model, he was asked to be a member. Though he dreaded sharing his discoveries with "the prejudic'd and censorious multitude," pride led him to accept the invitation. The rattling carriage journey from Cambridge to London would prevent Newton from attending the society's meetings but nothing could shield him from the backbiting that was common among boys of any age. Six weeks after he joined the society, he wrote to its secretary, Heinrich Oldenburg, offering his "Theory on Light and Colours."

On February 8, 1672, the next meeting of the Royal Society was called to order. First on the agenda was a report on how the moon affected a barometer. Next came a study of tarantula bites. Then Secretary Oldenburg presented the paper he had received that morning from the society's distinguished new colleague in Cambridge. Reports before the society usually began with gushing praise for fellow members or expressions of the author's humility, but Newton's paper cut through that fog like a beam from a lighthouse.

> 1. As the rays of light differ in degrees of Refrangibility, so they also differ in their disposition to exhibit this or that particular colour. Colours are not Qualifications of Light, derived from Refractions, or Reflections of natural Bodies (as 'tis generally believed,) but Original and connate properties, which in divers Rays are divers.
>
> 2. To the same degree of Refrangibility ever belongs the same colour, and to the same colour ever belongs the same degree of

> refrangibility. The *least Refrangible* Rays are all disposed to exhibit a *Red* colour . . . the most *refrangible Rays* . . . to exhibit a deep *Violet* colour.

Listening intently was Robert Boyle, whose color theory Newton had once believed. Fifteen years older than Newton and steeped in Descartes and Galileo, Boyle still held that white light was somehow modified into colors. Now here was this young upstart insisting that light was *made* of separate colors. Secretary Oldenburg read on:

> But the most surprising and wonderful composition was that of *Whiteness*. There is no one sort of Rays which alone can exhibit this. 'Tis ever compounded, and to its composition are requisite all the aforesaid primary colours, mixed in due proportion. Hence therefore it comes to pass that *Whiteness* is the usual colour of *Light*, for, Light is a confused aggregate of Rays indued with all sorts of Colours, as they are promiscuously darted from the various luminous bodies.

What Boyle thought about Newton's theory remains unknown, but elsewhere in the crowded room sat a short, hunched, conniving man who would soon make Newton's life miserable. Robert Hooke specialized in microscopes and had written the groundbreaking *Micrographica*, which Newton had admired. Hooke and Newton seemed born to be enemies: the one tall with aquiline features, the other stooped and gnarled; one reclusive, the other a regular of London coffeehouses; one a genius, the other wanting desperately to earn that accolade. Their thirty-year feud did not begin that February afternoon. Hooke sat politely as Newton's paper was read, then joined the unanimous vote that the theory be published lest it be stolen. But when the meeting adjourned, Hooke adjourned to nearby Joe's Coffee House, on Fleet Street just a few blocks from the Thames, to plot against his new rival.

Hooke, though no fan of the Frenchman Descartes, agreed that light was pressure, or as Thomas Hobbes had posited, a pulse. Colors, Hooke believed, were limited to two—white and black—the rest resulting from "an impression on the retina of an oblique and confuse'd pulse of light." The theory that color was white light split into component parts struck Hooke as absurd, and the spread of such nonsense seemed dangerous. As the Royal Society's Curator of Experiments, Hooke was assigned to write a response to

Newton's theory. After briefly reviewing the paper but doing none of the experiments, he began.

Back in Cambridge, Newton was delighted to learn that the Royal Society would publish his paper. A few days later, however, he read Hooke's response. Though praising "the Excellent Discourse of Mr. Newton," Hooke dismissed all the prism experiments as derivative, claiming to have done them years earlier. Likewise, Hooke scoffed at the suggestion that white light contained "divers" rays of primary colors. "Even those very experiments which he alledged," Hooke wrote, "doe seem to me to prove that light is nothing but a pulse or motion propagated through an homogeneous, uniform, and transparent medium." A prism might "split" light, but were the colors any more distinct than the mingled sounds of an orchestra? "Colour is nothing but the Disturbance of the light by the communication of that pulse to other transparent mediums . . . [and] the two colours (then which there are noe more uncompounded in Nature) are nothing but the effects of a compounded pulse or disturbed propagation of motion caused by Refraction." Mr. Newton, Hooke concluded, should spend his time improving his little telescope, much like the one Hooke now claimed to have made years earlier.

If rage had a color—bloodred, no doubt—Newton would have been as crimson as the draperies and furniture that always dominated his household décor. He somehow muted his anger, asking only that the Royal Society try his experiments for themselves. Hooke refused. Stiff exchanges continued throughout the spring, Hooke claiming to have seen it all before, Newton answering with more requests to simply *do* the experiments. Hooke refused. Newton soon reeled from further criticism. A prism, Astronomer Royal John Flamsteed assured the Royal Society, indeed scattered a single beam into separate colors, but each color blended with the others to "cause no determinate but a confused color mixed of all sorts." When the Royal Society's *Philosophical Transactions* published Newton's "Theory of Light and Colours," still more doubt came from Paris. Newton's oblong spectrum, a French skeptic wrote, was caused by light striking the prism from different parts of the sun. And colors, this disciple of Descartes argued, were just blends of black and white.

Newton spent the summer of 1672 in the countryside. He had made more optical discoveries but now he withheld them. His revolutionary theory about light, debated by just a handful of curious men, would not reach the rest of the world for another three decades. Hooke, ordered to reconsider Newton's original paper, suggested to his rival that they correspond in private. Newton agreed, but the feud was far from finished.

Further skepticism soon arrived. The Dutch astronomer Christiaan Huygens argued that light contained only two hues, but neither black nor white. "Neither do I see, why Mr. Newton doth not content himself with the two Colors, Yellow and Blew; for it will be much more easy to find an Hypothesis by Motion that may explicate these two differences, than for so many diversities as there are of other Colors." Newton replied that if the eminent astronomer wanted to paint the world in two colors, he should become an artist.

In March 1673, Newton asked to leave the Royal Society. Secretary Oldenburg ignored the request, but suddenly the society heard only silence from Mr. Newton. Alone again with his prisms, Newton spent the next two years, as the poet William Wordsworth would later imagine him, "voyaging through strange seas of Thought, alone." Though he was barely past thirty, his long hair was now gray and unkempt. He regretted ever having approached the Royal Society, ever letting his discoveries leave his chamber. "In hunting for a shadow hitherto," he wrote, "I had sacrificed my peace, a matter of real substance." He wrote no more letters to Robert Hooke, Christiaan Huygens, or anyone else who might care to discuss light. Working in his candlelit room, he filled notebooks with studies of matter and motion but also of alchemy and ancient civilizations. He strode the narrow pathways of Trinity College, brooding, pondering, sometimes stopping and stretching out a stick to sketch an equation in the gravel.

The feuds that sent Newton into seclusion transcended mere color. With the Scientific Revolution in full bloom, the essence of light was up for grabs. By the mid-1600s, though England's spreading Quaker sect was seeking God's "inner light," no philosopher-scientist thought light to be spiritual. It had to be *something*, be made of *something*, behave like *something*. But what? The devious Robert Hooke clung to his theory that light was a pulse. Others held with Descartes that it was pressure. But a new theory was dawning. Perhaps light did not behave like tennis balls or vibrations through a stick. Light's best model might be found at the edge of the ocean.

Back in 1661, when Newton was just starting his studies at Cambridge, a professor in Bologna had noticed yet another puzzle of light. Though he taught optics, Father Francesco Grimaldi had no illusions of explaining what light might be. "Let us be honest," the Jesuit father wrote, "we do not really know anything about the nature of light and it is dishonest to use big words which are meaningless." But in studying shadows, Father Grimaldi noticed something that had struck Leonardo da Vinci. "Shadow is divided into two parts," Leonardo had written, "of which the first is called primary shadow; the second, derivative shadow." Leonardo attributed derivative shadows—the fuzzy shade surrounding a darker umbra—to reflected light but Father Grimaldi detected another cause. Making a tiny hole in his shutter, he let a pencil beam enter. When he placed a coin in the beam, its shadow was fuzzy and slightly larger than it should have been. "Light," Grimaldi concluded, "is propagated or diffused not only directly and by refraction and by reflection, but also in a fourth way, by diffraction."

Diffraction, the slight bending of light as it streams past a hard edge, explained fuzzy shadows. It also explained why, when Grimaldi bounced light off a scratched metal plate, the beams scattered into faint colors. Grimaldi's etched plate was the first diffraction grating. These days diffraction gratings make rainbow patterns in toys and are an important part of holograms. Every DVD and compact disc diffracts light into colors, as do certain feathers. Corpuscles of light could not explain diffraction. Particles would stream past any edge without scattering around it. "The modification of light by means of which it becomes permanently and apparently colored," Grimaldi wrote, "can probably be said to be due to an undulation." Father Grimaldi died in 1663, before he could experiment further. Fellow Jesuits buried him in a tomb inscribed HE LIVED AMONG US WITHOUT ANY QUARREL. But Grimaldi's suggestion that light was an "undulation" started a quarrel that would last 250 years.

In the mid-1670s, one of Newton's many critics codified the emerging "wave theory" of light. Christiaan Huygens was a mathematician on a par with Archimedes, Descartes, and perhaps Newton. Pioneer of probability theory, discoverer of Saturn's rings, Huygens was working in Paris when he began his *Traité de la lumière* (Treatise on light). In just ninety pages, the treatise disputed all existing theories about what light might be. It could not be pressure, Huygens said, for what would happen to two people staring into each other's eyes? Wouldn't the light pressing in opposite directions collide and hamper their vision? Nor did light travel faster in water than in air, as both Descartes and Newton believed. Denser materials—glass, water, a prism—all slowed light, bending its rays. But were they actually rays, or even particles flowing in straight lines? Huygens maintained that light had to be a wave, like those approaching the shore. Each part of the wave touched another part, creating secondary wavelets flowing around a wave front. To predict how waves and wavelets would refract, Huygens calculated angles, analyzed sequential motions using adjacent triangles, and applied the

Snell-Descartes law. The result was a precise mathematical model of wave behavior easily applied to light. "We will cease to be astonished when considering that at a great distance from the luminous body an infinity of waves, even if emitted by various points of this body, join together forming one single wave which consequently is strong enough to be detected."

Huygens, like Newton, shied from publicity; he would not publish his treatise for another dozen years. By then, Newton had emerged from solitude to send his next submission to the Royal Society.

Throughout the Christmas season of 1675, meetings of the Royal Society refrained from torturing animals or studying deformities. Instead, members discussed Newton's "A Hypothesis Explaining the Properties of Light." Newton, as usual, was not present. Through several afternoons, members debated before by the glowing fire. Newton's latest paper showed the scars of previous criticism. Although he usually relied solely on empirical proof, now he labeled his discourse a hypothesis. The label freed him to speculate on what light might be, yet still, he tiptoed around his answer.

> Were I to assume an Hypothesis it should be this if propounded more generally, So as not to determine what Light is, farther then that it is something or other capable of exciting vibrations in the aether . . . Others may suppose it multitudes of unimaginable small & swift Corpuscles of various sizes, springing from shining bodies at great distances, one after another, but yet without any sensible interval of time, & continually urged forward by a Principle of motion . . . Some would readily grant this may be a Spiritual one; yet a mechanical one might be showne . . .

Newton's latest hypothesis set Robert Hooke and his friends to lengthy debate in London coffeehouses. The architect Christopher Wren joined the discussion in January, by which time Hooke's diary boasted of how he had "show'd that Mr. Newton had taken my hypothesis of the pulse or wave." But when Newton submitted more experiments, Hooke saw his own shade of red.

In his *Micrographica*, Hooke had described how thin transparent plates, when pressed together, displayed concentric circles, multicolored, with a black oval at the center. Hooke had merely observed these rings, but now Newton described them. Pressing two prisms together, Newton first saw the rings, eerie, shifting, seeming to float between glass. The central spot was surrounded by "many slender Arcs of Colours which at first were shaped almost like Conchoid." Using his own thin plates, Newton watched the rings expand when pressed, fade when touched by water. He sketched the rings, labeling each color circling the dark center. He calculated the refractions and measured the circumferences of what are now called Newton's Rings. Reading of Newton's experiments, Hooke erupted. Newton answered that he had credited Hooke with the discovery before making his own observations. "I suppose he will allow me to make use of what I tooke the pains to find out." And when Royal Society members continued to challenge his theories, Newton repeated his initial plea. *Do* the experiments. Finally, they did. If summer solstice at Stonehenge seems like "the sun's birthday," then the modern conception of light has its own date and place of birth.

London. Gresham College. April 27, 1676: All Royal Society members, except Newton, are in attendance. Outside, the spring blossoms. Inside, the stooped, diminutive Robert Hooke shines a beam through a shutter and aims it at a prism. A rectangle of brilliant colors shines on one wall. Hooke isolates a single color and beams it through a second prism. The red stays red. Hooke overlaps single hues to make different shades, then follows Newton's other instructions. According to the Royal Society record of that day, light behaved "according to Mr. Newton's directions, and succeeded, as he all along had asserted it would do."

In 1677, Newton began preparing his *Opticks* for publication, but a fire in his room at Trinity College burned many of his papers. Then Royal Society Secretary Heinrich Oldenburg died. In his place, members chose—Robert Hooke. Newton retreated, "for I see a man must either resolve to put out nothing new or become a slave to defend it." Over the next decade, he lectured to the walls, and wrote treatises on alchemy and the Temple of Solomon, whom Newton called the world's greatest philosopher. Living alone, he studied gravity and motion, and fine-tuned the most far-reaching creation of his annus mirabilis—the calculus. (The German philosopher Gottfried Leibniz created his own version of the calculus at about the same time, but Newton published his version in more detail and got the immediate credit.) In 1687, Newton's universe, except for his work on light, emerged in the *Principia*. Written in Latin and bursting with equations, postulates, problems, and theorems, this universe bound in parchment was translated into many languages, making Newton famous far beyond light's small cadre of students. A French philosopher asked about Newton: "Does he eat & drink & sleep? Is he like other men?" A Scottish mathematician thanked

Newton "for having been at pains to teach the world that which I never expected any man should have known." And in 1698, when Russia's Peter the Great came to London, he asked to see three wonders—British shipyards, the Royal Mint, and Isaac Newton.

As his fame spread, Newton was urged to publish his optical discoveries. "You say, you dare not yet publish it," a mathematician wrote him. "And why not yet? Or, if not now, when then? . . . Meanwhile, you loose the Reputation of it, and we the Benefit." Newton responded with silence. Finally, in 1703, Robert Hooke died. By unanimous vote, Royal Society members chose Newton to succeed him. Study and seclusion had by then turned the "poore . . . paile . . . little fellow" into the man one colleague called, "the most fearful, cautious, and suspicious Temper that I ever knew." But few refuted his theories, and with Hooke gone, Newton finally let the world know what he had discovered so long ago about light.

Opticks was published in 1704. Its experiments were models of precision, yet tasked with describing light's quirkiest behaviors, Newton struggled. Explaining why water reflects some particles of light and refracts others, he wrote of "fits"—"fits of easy Reflexion" or "fits of easy transmission." Absorbed light was "stifled or lost," excess illumination was "foreign light," while its common properties were merely "vulgar." But despite its awkward prose, *Opticks* enhanced understanding of light as dramatically as any prism bent its rays.

The book's first two sections described the experiments Newton had conducted decades earlier. Prisms and rings. Paging through *Opticks*, the reader saw light filtered through hole F, beamed at prism ABC, streamed through lens MN, emerging in separate beams P, R, and T. Propositions grew into problems and problems into theorems. Newton claimed he was writing "an Introduction to Readers of quick Wit and good Understanding not yet versed in Opticks," but the book required a thorough understanding of Cartesian geometry. In the third book, Newton tried to replicate the diffraction discovered by Father Grimaldi, whom he called Grimaldo. Like the Jesuit father, Newton used the thinnest possible light—he estimated his beam at 1/42 of an inch. He aimed it at a single hair. He shone it between two knife blades. He moved white paper closer, farther. Though he saw colored bands and two beams that shot "like the Tails of Comets," he drew no conclusions about this fourth way light traveled. He had planned a full study of diffraction but could not define it with the certainty his exacting mind demanded. Instead, he closed his treatise with queries, guides to further study.

The twenty-eight queries in *Opticks* reveal how light puzzled even Isaac

Newton. Of ether he speculated about a "much subtiler Medium than Air," but admitted, "I do not know what this Aether is." Paying homage to wave theory, he asked, "Are not the Rays of Light in passing by the edges and sides of Bodies, bent several times backwards and forwards, with a motion like that of an Eel?" Because the speed of light had recently been estimated by a Dutch astronomer charting the moons of Jupiter, Newton ventured his own guess—196,000 miles per second, bringing sunlight to earth in seven minutes. He was just one minute and 10,000 miles per second off. Finally, the Newton no one knew—the alchemist, the mystic, the spiritual seeker—probed the very nature of light. "Are not the Rays of Light very small Bodies emitted from shining Substances?" And with change as the eternal nature of nature, turning tadpoles into frogs and worms into flies, "why may not Nature change Bodies into Light, and Light into Bodies?" Alas, Newton concluded, humanity might already have answered these questions but for precious time wasted in "the Worship of false Gods . . . the Sun and Moon, and dead Heroes." In subsequent editions of *Opticks* Newton added more queries, yet in each version his experiments remained intact, inspiring the growing numbers of the curious.

Common wisdom and the poet Alexander Pope held that once Newton published *Opticks*, "all was light." But scientific revolutions do not storm ideological barricades. Instead, as the philosopher Thomas Kuhn noted, new theories advance in increments, leading to feuds, wars of words, and finally, a paradigm shift. In 1704, the Royal Society met Newton's *Opticks* with silence; this was old news stirring bitter memories. "The book," John Flamsteed wrote, "makes no Noyse in Town as the *Principia* did." The Royal Society's Parisian counterpart, l'Académie des Sciences, listened to Newton's treatise over ten sessions but still trusted (ahem!) Monsieur Descartes. A few Frenchmen replicated Newton's prism experiments, but others fumbled with them and saw no clear proof. Skepticism survived into the 1720s. By then, Sir Isaac Newton was in his eighties, suffering from kidney stones, withering away. When he died, in 1727, he was given a full state funeral, the first commoner to receive one. Among those who watched Newton's coffin interned beneath the Gothic light of Westminster Abbey was a French writer in exile who went by the name of Voltaire. He wrote home: "I have seen a professor of mathematics, simply because he was great in his vocation, buried like a king who had been good to his subjects."

When Voltaire returned to France, he met the elegant Marquise du Châtelet. The marquise was a mathematical prodigy, but Voltaire was not intimidated. He and the marquise, soon living together, began teaching

Newton to the rest of Europe. She translated the *Principia* into French while Voltaire, in his *Letters on England*, simply swooned over Newton, "whose equal is hardly found in a thousand years." Noting that Newton had dissected light "with more dexterity than the ablest artist dissects a human body," Voltaire proclaimed, "This man is come." *Letters on England*, critical of the French, was banned in Paris. Voltaire went into exile again, but when reunited with the marquise he began a book on Newton. Now for the first time since creation, light would be explained neither by holy mystics nor long-winded wizards but by the most gifted writer of the age.

Aiming at the widest possible audience, Voltaire patiently reviewed the canon of optics from Ptolemy's coin-in-a-cup to Grimaldi's diffraction. Light, Voltaire proclaimed, was "fire itself," projected in particles "thrown from the Sun to us, and as far as Saturn, &c, with a Rapidity that amazes the Imagination." (Voltaire guessed that light traveled 1.66 million times faster than a cannonball.) He explained particle theory, positing that each particle of light had a fixed weight, red being the heaviest and thus the most refracted. He explored rainbows and the wonders of refracted light, which he suggested readers observe through their own prisms, as he had. And Voltaire scoffed at those who still sided with Monsieur Descartes. "Is it because they are born in France that they are ashamed of receiving truth at the hands of an Englishman?"

In 1738, *The Elements of Newton's Philosophy Within Reach of Everyone* was published in Paris. And the paradigm shifted. "All Paris resounds with Newton," one observer wrote, "all Paris stammers Newton, all Paris studies and learns Newton." A friend of Voltaire's soon introduced Italy to Newton, writing *Il Newtonianismo per le dame* (Newton for women). Throughout the rest of the eighteenth century, across the continent, Newton's was the light that prevailed, the light that refracted, the light that behaved precisely as he said it would. *Opticks* was taught in universities, celebrated in poems, hailed in royal courts. Its basic assertion—that white light was made of separate, immutable colors—became everyday knowledge. Any fool with a prism could know light as Isaac Newton had. Voltaire: "There is a new universe opening itself to the eyes of those who choose to see it."

Alone in his dark chamber, he places a prism "at the Hole of the Window-shut . . . so that its Axis might be parallel to the Axis of the world." Against the opposite wall, he sets an open book. Between prism and book, he fixes a lens. He adjusts the prism to send light's colors through the lens, then focuses a single red beam on a printed page. He stands, letting the sun unveil the silent show. "And then I stay'd till by the Motion of the Sun, and consequent Motion of his Image on the Book, all the Colours from that red to the middle of the blue pass'd over those Letters."

CHAPTER 9

"A Wild and Harmonized Tune": The Romantics and the Light Seductive

Hear the voice of the Bard,
Who present, past, and future sees;
Whose ears have heard
The Holy Word
That walked among the ancient trees;

Calling the lapsed soul,
and weeping in the dew;
That might control
the starry pole,
And fallen, fallen light renew!

—WILLIAM BLAKE, "HEAR THE VOICE"

Like snow falling and clouds adrift, light beguiles us with silence. The most livid sunrise unfolds in muted glory. The wheel of constellations that spins through the night makes no sound. Not even the bold streak of a meteor disturbs the stillness. Only within us does the music of celestial light play on. As if scoring our own movie, we watch sun and moon with concertos in our heads. Sunrise is Beethoven, sunset Mozart. When light first became music, however, no one in the Vienna concert hall was prepared, never having heard it make a sound. The year was 1798, the dawn of the Romantic era.

Blazing like a flare over a battlefield, the Romantic era rebelled against all that was rational, analytical, or labeled "the Enlightenment." Romantics, from Byron to Goethe, from Chopin to Keats, rejected the empiricism of Newton, the "social contract" of Rousseau, and the "pure reason" of Immanuel Kant. Romantics championed the individual over the industrializing masses and preferred heartfelt emotion to "meddling intellect." Above all else, they cherished the elusive mix of beauty and longing known as "the sublime." Romantics fell in love over and over, and drank and wrote and drank and painted and drank and composed through affairs that were bound to end badly. When they could not find romance in art or in a beloved's eyes, they sought it in nature. Long walks through moonlit graveyards or along Italian beaches left them enraptured by wind, waves, and light.

The Romantics made Isaac Newton their nemesis. Throughout the eighteenth century, Newton's legacy had verged on legend. Cited by scholars yet familiar to all, Newton was the Enlightenment's secular deity—"our philosophic sun," "the godlike man," "who was himself the light." In 1784, a French architect planned a model city featuring a Newton cenotaph, 750 feet in diameter, mounted on a pedestal bathed in light. Though the cenotaph was never built, Newton's light remained ascendant. Students, not just of physics but of art and literature, were expected to know the basics of light, how it refracted and reflected, and what it might be made of. Writers used light in popular novels. On his celebrated travels, Lemuel Gulliver met the Laputans, people who lived for light, studying the stars and starting each morning by checking the sun's health. Artists paid tribute to Newton in elaborate color schemes—wheels, triangles, and pyramids structuring the colors of his spectrum. When the French chemist Antoine Lavoisier listed natural elements, thirty-three in all, the first was not hydrogen, as the periodic table would later establish, but light.

Throughout the Enlightenment, philosophers and astronomers expanded the cosmos. In 1755, Immanuel Kant was the first to speculate on the vastness of the night sky. "With what astonishment are we transported when we behold the infinite multitude of worlds and systems which fill the extension of the Milky Way!" Kant wrote. "There is here no end but an abyss of a real immensity, in presence of which all the capability of human conception sinks exhausted." A generation later, the astronomer William Herschel discovered a planet too distant to be seen with the naked eye, Uranus. Working with his sister Caroline, Herschel went on to discover galaxies—hundreds of them. Beyond the Milky Way, Herschel said, lay not the earth's shadow hiding heaven's light, but an immense void dappled by "the chaotic material of future suns." Light was the essence of stars, and the sun just one among billions. This was part of the "new universe" Voltaire had glimpsed through Newton's prism. Some, however, refused to see it.

A generation of Romantics revered light as more than a pencil beam piercing a dark chamber, more even than the stuff of stars. Perhaps it was no longer God, nor even his self-portrait, yet Romantics saw light as the essence of beauty and transcendence. If the tools of science—mirrors and prisms and equations—had robbed "fallen light" of its grandeur, then the tools of art would restore it. And throughout the Romantic era, light was eulogized in verse and spun into symphony. The sun blazed more brightly than ever on canvas, and the moon became a matchmaker. Light's Romantic era began with the blare of an orchestra.

The novelty of an oratorio entitled *The Creation* aroused interest but it was the composer who drew the overflow crowd to Vienna's Schwarzenberg Palace on a spring night in 1798. Crowds teemed outside the snow-white Baroque edifice, forcing police to scatter onlookers who just wanted to see their beloved "Papa Haydn." The son of an Austrian wheelwright who had made music a daily presence in the Haydn family cottage, Joseph Haydn had risen to the courts of kings and princes. Royal commissions had supported most of his works, including concertos, oratorios, and more than a hundred symphonies. In his mid-sixties, still one of Europe's most celebrated composers, Haydn began composing *The Creation.*

The idea came from an English libretto given him by a friend in London. The libretto, based on Genesis and Milton's *Paradise Lost,* detailed the biblical creation, even opening with "In the beginning . . ." Haydn was intrigued, especially after meeting the astronomer William Herschel, who told him of the vast, light-filled universe. Back in Vienna, Haydn had a friend translate the libretto into German, then turned to prayer. "Never was I so devout as when composing 'The Creation,' " he later wrote. "I knelt down every day and prayed to God to strengthen me for my work."

On April 20, the elites of Vienna, in billowing gowns, high collars, and powdered wigs, gathered inside the palace concert hall. Chandeliers glittered; jewels sparkled on bare necks and bosoms. The orchestra took the stage, followed by three soloists, a small chorus, and finally, the plump, white-wigged Haydn, baton in hand. One in the crowd recalled the moment: "The most profound silence, the most scrupulous attention, a sentiment, I might almost say, of religious respect prevailed when the first stroke of the bow was given." *The Creation* began with muted trumpets flaring, then fading into primeval Chaos. In a haunting minor key, violins, flutes, and clarinets roamed through dissonant notes, swirling as if "without form and void." The audience sat as Chaos continued, rising, falling, flowing. Then, at the sounding of three ponderous notes, the hall fell mute. A man in formal attire stepped to the front of the stage. Straightening himself, he began in a monotonal bass:

Im Anfange schuf Gott Himmel und Erde . . .

Violins and a clarinet filled in. The bass continued:

und die Erde war ohne Form und leer . . .

Four female voices rose like the spirit of God "upon the face of the waters." The sopranos stilled, then softly sang God's first words:

Es werde Liii—cht.

Three violins plucked a velvet pizzicato note. Silence again. The audience sat, hushed. Then the chorus sang sotto voce,

Und es war . . . ("and there was . . .")

Two seconds passed as the entire orchestra drew breath. Then . . .

Trumpets blasted, bassoons blared, tympanis boomed, oboes flared, cornets chimed in, all exploding on a single note. The lilt of strings mushroomed like billowing clouds. The crescendo rose and swelled. At the podium, Haydn flailed his arms, sending the violins still higher, cuing the ultimate blast. A friend later recalled, "When light broke out for the first time, one would have said that rays darted from the composer's burning eyes." After fifteen seconds, trumpets beamed their last rays, and the soloist sang of God seeing the light, "*dass es gut war*" ("that it was good"). The audience struggled for calm. Light, First Light, had resounded. Several minutes passed before the excitement settled and a new aria could begin.

Over the next decade, Haydn's *Creation* became the most discussed musical work in Europe. In concert halls from London to St. Petersburg, audiences *heard* light. The oratorio delighted Napoleon in Paris and employed Beethoven on piano in Vienna. Light, the Romantic, had begun its seduction. The next to succumb was a painter.

In the 130 years since the death of Rembrandt, light had dazzled few canvases. Fruit and vases still gleamed, yet Baroque and Mannerist painters preferred burnt umber backdrops, muted light, and no artist seemed especially captivated by radiance. Then the same year Haydn put light to music, a short, homely, reticent painter with no family on whom to bestow his affections fell in love.

As a child raised in the smoke and soot of industrial London, Joseph Mallord William Turner became fascinated by the sun. By 1775, the year of Turner's birth, daylight barely dawned over the Thames. Choked by chimneys and mills, the London sky resembled, as the painter John Constable remembered, "a pearl through a burnt glass." Turner, the son of a poor barber and a mother flirting with madness, was drawn to light, any light that cut through the gloaming. Though not religious, he was already moving toward the conviction he would express shortly before his death, that "the sun is God." His talent was noticed in his teens, and by his early twenties he was being compared to Rembrandt. But like Rembrandt's early paintings, Turner's displayed no special light. His landscapes were brighter than most, yet their radiance was spread evenly over meadows, manors, and cathedrals. This ordinary light also reflected Turner's character. Of all light's disciples—the *phusikoi*, the monks, the poets, the eccentrics—Turner was surely the dullest. Shunning human contact, smiling only when surrounded by children, he struck everyone as peculiarly mundane. "It seemed to amuse him," the *London Times* wrote, "to be but half understood." Turner wrote no memorable letters, made only the hastiest notes about art, and was consumed solely by a compulsion to sketch whatever he had traveled far to find. "The man must be loved for his works," an acquaintance noted, "for his person is not striking, nor his conversation brilliant."

Turner was already making a good living from landscapes when, in 1798, he began to paint the embattled sunlight of his boyhood. Poetry was one prompt. London's Royal Academy of Arts, where Turner had studied and would soon teach, had begun posting inspirational verse alongside paintings hung in its gallery. This change, reflecting the academy's belief that poetry and painting were "sister arts," moved Turner to scour the works of Milton. In *Paradise Lost* he read:

Hail, holy Light, offspring of Heaven first-born!
Or of th' Eternal coeternal beam
May I express thee unblamed? since God is Light . . .

For Turner, Milton was a revelation. Perhaps others also dared to consider the sun a deity. Soon a wealthy patron invited Turner to his home to admire a painting. The work, by the French artist Claude Lorrain, depicted a seaport before a dazzling sunset. As he absorbed this remarkable light, Turner burst into tears.

"Why are you crying like that, my boy?" Turner's patron asked.

"Because I shall never be able to paint anything like that picture."

Within a year, Turner was challenging all the old rules. Ever since Leonardo, artists had seen the sky as blue. Landscape after landscape showed the same pale blue sky, the same fluffy clouds, the same ho-hum light. Blue, however, was not the color of the London skies Turner remembered, nor the color he envisioned if he could just clear away the smoke and soot. Turner's light, the light already creeping into his works, the light that would speak to anyone ever stunned by a sunset, was *yellow*. In 1799, Turner finished his watercolor *Caernarvon Castle*. Suddenly, as if the sun had blasted Newton's prism across his chamber, light on canvas burst through all boundaries. In this first Turner recognizable as "a Turner," the old Welsh castle is backlit by a searing sun. Dead center on the small canvas, the tuppence disc melts into a simmering sky. *Caernarvon Castle* burns with a heat that singes the air and ripples the river below. This was more than painterly light; this was the light of an artist unafraid to worship the sun, who suffered both its joy and its absence. And from here onward, Turner's light brightened, and his darks darkened, though rarely in the same painting. Other artists might prefer Leonardo's advice, but Turner had read Giovanni Paolo Lomazzo, the same theorist who had inspired Caravaggio. In one of Turner's sketchbooks he scrawled an homage:

Light
Primary Light that which is received direct . . .

Lomazzo gives this to emanation of the Deity or adoration of Glory.

On into the 1800s, Turner continued to sell subdued renderings of old abbeys and churches, but in paintings made not for money but for love, he amped up his light. His seascapes featured roiling waves streaked by the light of salvation. His light over England's pastures was luminous yellow, as if there were no blue sky anymore. And his beloved sun, as in one title, *Rising Through Vapor*, scorched the eyes.

Like his fellow Romantics, Turner loved distant lands as much as he loved any woman. Never marrying, he lived intermittently with one widow, then decades later, another. Though rumored to have two daughters and a son with the first, he often said his paintings were his children. Relentless wandering took him throughout the British Isles, to Paris and the Alps, then home with hundreds of sketches to inspire increasingly wild works, all painted in the studio. Once, at his window overlooking the Thames, he gestured to the horizon. "Have you seen my study?" he asked a friend. "Sky and water. Are they not glorious? Here I have my lessons day and night."

In 1812, Turner took another leap in *Snow Storm: Hannibal and His Army Crossing the Alps*. Here, tiny forms trudge beneath sooty clouds that swirl like ocean waves. The sun is an egg yolk edged in black; its light, like the godlight of Mani and Zoroaster, menaced by darkness. For the next several years, the sun made cameo appearances in Turner's commercial work, yet the demands of the staid London art world kept its brilliance at bay. Then in 1819, Turner traveled to Venice.

Anyone who lives near an ocean or river knows that light and water complement each other in charming ways. Water turns sunlight into sequins and tints the horizon above the Hudson, the Arno, the Seine, with a palette more resplendent than any over ordinary land. Artists in the Netherlands have long spoken of "Dutch light," the special brilliance of skies above Vermeer's Delft and its surroundings. Dutch light radiates off the abundant waters of the Low Countries; Turner saw a similar glow over the canals of Venice. He made no paintings there, just sketches, before going on to Rome, yet he returned from Italy with its afterglow inside him. Within a year, his light was unlike any ever seen on canvas.

In *Venice, Looking East from the Giudecca: Sunrise*, a platinum-blond horizon floods a faint skyline of cupola and campanile. Another of Turner's Venice paintings smears the sun across a vanilla backdrop. Elsewhere, Turner's moonlight commands ocean waves while his daylight consumes the atmosphere. For the rest of his life, Turner would traffic in light. More radiant than Rembrandt's, Turner's is primal light. The critic William Hazlitt wrote that Turner "delights to go back to the first chaos of the world, or to that state of things when the waters were separated from dry land, and light from darkness . . . All is 'without form and void.'" To stare into a later Turner is to drift toward the sun itself, to rise like Dante toward paradise, moving closer, closer to the source, while its music swells within.

Turner's pure light was not the product of pure paint. Unlike Rembrandt, Turner employed many tricks. Whereas the Dutch master and most other painters preferred an ocher background, Turner prepared his canvases with a white base. To make his oils as luminescent as his watercolors, he then layered on a shiny veneer. The final work was glazed and scumbled—softened with a semiopaque overlay. Turner kept one fingernail long and used it to attack his canvases, spreading and cutting, sometimes even spitting into the paint. He applied stale bread crumbs to highlight certain hues and was once seen mixing powdered pigment with stale beer. Enchanted by color, Turner adopted the latest available shades—cobalt blue in the 1810s, emerald green and French ultramarine a decade later. But it was yellow that made a Turner "a Turner," yellow seeping into white just as sunlight dissolves into a thinning sky. John Constable called Turner's colors "tinted steam," and by the 1840s, when Turner was nearing seventy, he was known as "the painter of light." Often his light churned through massive storms or waves that presaged Jackson Pollock's proclamation a century on: "I am nature." Turner's last canvases were washed in eddies of color, suggesting the raw energy physicists would soon discover light to be.

Critics who had compared him to Rembrandt were angry and confused. Turner's light had become too raw for the emerging Victorian age. The humor magazine *Punch* suggested an all-purpose Turner title: "A Typhoon bursting in a Simoom over the Whirlpool of a Maelstrom, Norway; with a ship on fire, an eclipse, and the effects of a lunar rainbow." Queen Victoria refused to bestow a knighthood on Turner, thinking him quite mad. Only the young critic John Ruskin, whom Turner befriended, came to his defense. "Why then do you blame Turner because he dazzles you?" Ruskin wrote. "If . . . you had paused but so much as one quarter of an hour before the picture, you would have found the sense of air and space blended with every line, and breathing in every cloud, and every colour instinct and radiant with visible, glowing, absorbing light."

England continued to scoff, but in 1870, two young French painters visiting London took notice. Their names were Camille Pissarro and Claude Monet. By then, Turner had been granted his final wish. In 1855, confined to a wheelchair in his studio on the Thames, he asked to be moved to the window. There he saw his god, and there he passed into the light.

Putting light back on its pedestal was enough for most Romantics. Some, however, did not just disagree with Newton; they loathed him. For William Blake, Newton was the enemy of all things spiritual.

Reason and Newton, they are quite two things
For so the Swallow & the Sparrow sings

Reason says Miracle. Newton says Doubt
Aye that's the way to make all Nature out.

The mystical Blake, like Paul on the road to Damascus, saw photisms. Paul saw the glowing light of Jesus, but Blake, starting from his first vision at age ten, was visited by luminous angels. No one was going to tell him that light was not holy, and he missed few opportunities to vilify the "Newtonian phantasm . . . this impossible absurdity." Blake painted Newton as a soulless mathematician kneeling before creation, compass in hand, reducing all beauty to numbers. Newton also angered other poets. One inebriated evening at a dinner in London, Wordsworth, John Keats, and others began chiding their host, a painter, for adding Newton's image to a recent canvas. Keats, who had just finished a four-thousand-line poem about the moon, blamed Newton for reducing the rainbow "to a prism." Newton, another said, was "a fellow who believed nothing unless it was as clear as the three sides of a triangle." The drinking and damnation continued. Finally, all present offered an ironic toast "to Newton's health, and confusion to mathematics." Meanwhile in Germany, the poet Friedrich Schiller complained that Newton had turned the sun into a "soulless fireball." Schiller's friend went further.

Johann Wolfgang von Goethe, born in 1749, was a product of the Enlightenment yet became a founding father of Romanticism. His novel *The Sorrows of Young Werther* caused lovesick men throughout Europe to consider suicide. Goethe's plays were widely performed, his poems were set to music by Beethoven and Mozart, and his company was sought by anyone of note nearing his home in Weimar. Less attention, however, was paid to Goethe's studies of bones, plants, and geology. As perhaps the last intellectual who dared to span the widening gap between science and the humanities, Goethe felt compelled to study light. He began by blasting Newton.

Newton could only be challenged by a thinker renowned by all, and Goethe, feisty, original, and supremely self-confident, was such a thinker. Once the great Newton spoke, Goethe lamented, no one cared "if there are in this world painters, dyers, someone who observes the atmosphere and the colorful world with the freedom of a physicist, or a pretty girl, adorning herself according to her complexion." In contrast, he added, fellow Romantics "reserve the right to marvel at color's occurrences and meanings, to admire and, if possible, to uncover color's secrets." In 1788, after returning from a tour of Italy, Goethe began to counter Newton with his own theory of light. As it had through the pen of Voltaire, light again commanded the attention of the age's most gifted writer.

In Italy Goethe had visited several artists' studios. Having painted a little, he understood shadow and perspective, but "when it came to color; it seemed that all was left to chance." Goethe returned to Germany determined to study color. "Like the entire world, I was convinced that all colors are contained in light," he recalled. To test Newton's theory Goethe borrowed some prisms, but before he could make his own experiments, he was asked to return them. He was about to comply when he decided to look through a prism, something he hadn't done since boyhood. Holding a prism against his eyes, Goethe turned to a white wall expecting the glass to spread its light into the familiar colors. To his amazement, he saw the wall as . . . white. Goethe turned to the window. Again, the light was untinted. Yet along the horizontal frame, Goethe saw colored stripes, not Newton's spectrum but thin bands—orange above yellow at the window's upper edge and turquoise over a royal blue along the bottom. "I immediately spoke out loud to myself, through instinct, that Newtonian theory was erroneous. There was then no longer any thought of returning the prisms."

For the next two decades, with time out to write *Faust* and other world masterpieces, Goethe studied color. Newton had shone light through a prism, then two, but Goethe insisted this was not enough. Inspired by his Eureka moment, the German Romantic began peering through his glass at geometric patterns of black and white. Again he saw what Newton never noticed. Try it yourself.

Borrow a prism and promise to return it. Now put this book in your lap and turn to the black and white images (on page 120). Hold the prism to your eyes and study the rectangles. Behold the surprise Goethe saw. The black box above the white is topped by blue and violet bands while the border below is underlined in orange and yellow. Now look at the white rectangle contained inside the black box. The white is filled with color, an orange stripe fading to yellow on top, a blue stripe at the bottom. The adjacent black rectangle in a white box does the opposite. Puzzled and fascinated, Goethe played with patterns as Newton had with prisms, then rushed to friends to share the wonder. "I have yet to find someone who, upon conducting this last experiment, was not astounded," Goethe wrote. Clearly, such colors did not fit the theories of Sir Isaac Newton.

Goethe acknowledged no boundary between poetry and science. Instead, he pursued a synthesis—a "delicate empiricism." Goethe's light mirrored the ancient light of China's Taoism, not separate from the observer but fully integrated into human perception. Only light as interpreted by the human eye, understood by the human soul, deserved to be studied, Goethe argued. Beams isolated in experiments proved nothing. Even the most impartial experimenter could not help seeing what he wanted to see. "It is a calamity that the use of experiment has severed nature from man," Goethe wrote, "so that he is content to understand nature merely through what artificial instruments reveal . . . Microscopes and telescopes, in actual fact, confuse man's innate clarity of mind." Goethe was building a delicate bridge, a holistic approach some still hope will soften Western science. But it was easy to take this bridge too far.

In 1810, Goethe published *Zur Farbenlehre* (*Theory of Colors*). Running to four volumes, fifteen hundred pages, with hundreds of diagrams, *Zur Farbenlehre* veered from optics into psychology and speculation. The book, one scholar noted, was "not so much a study of color as a theology of color." Claiming that an observer "does not see a pure phenomenon with his eyes but more with his soul," Goethe studied after-images, optical illusions, colored shadows, and rainbows in soap bubbles. These colors—*all* color, he argued—were caused by horizontal borders between black and white. "There arises no prismatic color phenomenon unless an image is displaced, and there can be no image without a boundary." Where Newton saw seven inseparable colors, Goethe saw just three—the yellow, orange, and blue his prism had revealed. The rest, he maintained, were blends of light and dark. White light, he argued, was no composite but "the simplest, most homogenous, undivided entity that we know." Newton's assertion to the contrary "must move anyone who is not depraved to astonishment, indeed to horror."

Then Goethe waxed romantic. Each color, he theorized, created a mood. Yellow: "a serene, gay, softly exciting character." Blue: "a kind of contradiction between excitement and repose." Red: "conveys an impression of gravity and dignity, and at the same time of grace and attractiveness." Green: "The eye experiences a distinctly grateful impression from this color." Moving from the personal to the political, Goethe saw color in national character. "Lively nations, the French for instance, love intense colors, especially on the active side; sedate nations, like the English and Germans, wear straw-colored or leather-colored yellow accompanied with dark blue." Color perception must also vary by gender and age, Goethe went on, because "the female sex in youth is attached to rose color and sea green, in age to violet and dark green."

Like Newton, Goethe created a color wheel, but enhanced it with a poet's perception. Goethe's wheel set purple—a color for "monarchs, scholars, philosophers"—opposite yellow, the color of "bons vivants." Blue, the preferred hue of "orators, Historians, Teachers," lay across the wheel from orange, favored by heroes and despots. Green was for "lovers and poets." Red did not make the cut.

To the end of his life, Goethe considered his color theory the equal of his literary masterpieces. The rest of the world disagreed. Goethe's *Farbenlehre* was ignored by England's Royal Society and France's Académie des Sciences. Goethe denounced these esteemed bodies as "guilds," noting, "I have quarreled with the most excellent men on account of the disputed points in the *Farbenlehre*." Even Goethe's friends were skeptical. The budding philosopher Arthur Schopenhauer initially backed Goethe's theory but soon changed his mind. One day, when the two were discussing light, the dour Schopenhauer proposed that light might mimic the proverbial tree falling in the forest, making no sound. Goethe erupted: "What? Light should only exist inasmuch as it is seen? No! You would not exist if the light did not see you!"

Fellow Romantics found more in Goethe's theory. Beethoven considered *Zur Farbenlehre* more interesting than Goethe's later poetry. Turner subtitled one of his later paintings, *Goethe's Theory of Color*. In the twentieth century, Goethe's staunchest defender was the Austrian educator Rudolf Steiner. "Color is the soul of nature," Steiner wrote, "and when we experience color we participate in this soul." Today's Waldorf Schools, based on Steiner's theories, apply Goethe's mood-based colors to each classroom, grade by grade.

In challenging Newton, Goethe codified Romantic light. Surely, he suggested, we both see and *feel* light. The most romantic of the Romantics, however, needed no theory to prove that light was soulful. The moon was enough.

Before the Romantic poets made it their leading light, the moon had inspired worlds of myth and superstition. Aside from the broad palette of stars, the moon alone fought darkness, and its serene presence made it many things to many cultures. The moon was a calendar for farmers, a guide for travelers, and the goddess with a thousand names—Selene (Greek), Soma (Indian), Diana (Roman), Osiris (Egyptian), Inanna (Sumerian), Pe (Pygmy), Tsukiyomi (Japanese) . . . The moon symbolized eternity, fertility, and, as the adjective "lunar" suggests, lunacy. The Hindu Rig Veda spoke of "the mansion of the Moon," while the Qur'an, whose calendar is based on the lunar cycle, told of the prophet Muhammad splitting the moon in two. When not provoking madness or toying with fate, moonlight was compared to one animal after another—a bear, a frog, a rabbit, a bull. The moon was a hunter, a spinner, a weaver . . .

Early students had puzzled over the composition and influence of moonlight. Purple-robed Empedocles thought the "gentle moon" to be "air cut off by the fire" and frozen like hail. Aristotle linked waning moonlight to female menstruation and the full moon to infant mortality. Only when Galileo aimed his telescope at the lunar surface did the moon begin to lose the aura of folklore. The Romantics would restore that aura and add their own, making the moon a lamp for lovers.

India's erotic guidebook, *The Kama Sutra*, mentions the moon only as a diversion that lovers can admire *after*. The Italian Renaissance bard Petrarch wrote hundreds of sonnets to his beloved Laura, but few mention the moon. Shakespeare's sonnets, likewise, refuse to romanticize the moon. Even Juliet, when Romeo swears his love "by yonder blessed moon," rejects the metaphor:

> O, swear not by the moon, the inconstant moon,
> That monthly changes in her circled orb.

Full, waxing, or waning, the moon shone on human affairs from creation onward, yet until the Romantic era, few linked it with love, or dared to rhyme it with June.

Byron called his fellow Romantic poets "moon-struck bards," and each seemed to nurture a personal relationship with the moon. The intimate connection began with Goethe, whose tragic young Werther dreamed of wandering "across the heath in the pale moonlight." The Italian poet Giacomo Leopardi addressed several poems to the "gracious moon . . . O moon of my delight." Chopin's nocturnes filtered like moonlight through the trees. But of all the Romantics, the mooniest bunch lived in England.

In his autobiographical poem, "The Prelude," Wordsworth wrote:

The moon to me was dear;
For I could dream away my purposes,
Standing to gaze upon her while she hung
Midway between the hills . . .

Shelley was equally riveted by the moon, "that orbed maiden." In Italy he wrote letters by a moonlight "passing strange and wonderful," so entranced that his neighbors "believe I worship the moon." Shelley's moon, most often waning, was "like a dying lady, lean and pale / who totters forth, wrapped in a gauzy veil." Wordsworth and Shelley eulogized the moon in a few short poems, but Keats swooned again and again.

The young, frail John Keats spent many evenings on the cliffs of Margate, stunned into silence, watching the moon rise over the glittering North Sea. In verse and letters, Keats fawned over the "splendid moon," "glowing moon," "soft moon," "cheerful moon," "golden moon," "my silver moon." Finally, in 1818, he turned a Greek myth into his epic, *Endymion: A Poetic Romance.* As in the myth, the moon goddess falls in love with the sleeping shepherd boy Endymion. He is equally enamored:

What is there in thee, Moon! That thou shouldst move
My heart so potently? When yet a child
I oft have dried my tears when thou hast smil'd . . .

As the poem unfolds, the moon begs Zeus to make the shepherd boy sleep forever so she can dote on his blessed face. Endymion continues his paean:

And as I grew in years, still didst thou blend
With all my ardors: thou wast the deep glen;
Thou wast the mountain-top—the sage's pen—
The poet's harp—the voice of friends—the sun;
Thou wast the river—thou wast glory won;
Thou wast my clarion's blast—thou wast my steed—
My goblet full of wine—my topmost deed:—
Thou wast the charm of women, lovely Moon!
O what a wild and harmonized tune . . .

But loving the moon is not the same as love beneath the moon's light. Keats had made "the charm of women" just one of the moon's many roles, but Byron made romance the quintessence of moonlight.

The rogue adventurer and legendary lover Byron rarely watched the moon rise from seaside cliffs nor did he write letters by its light. One imagines he had other pursuits by moonlight. Having scandalized England with his affairs, Byron knew the moon's effect on lovers. Again and again in his masterwork, *Don Juan*, the rising moon leads the title character to seduction:

They look'd up to the sky, whose floating glow
Spread like a rosy ocean, vast and bright;
They gazed upon the glittering sea below,
Whence the broad moon rose circling into sight;
They heard the wave's splash, and the wind so low,
And saw each other's dark eyes darting light
Into each other—and beholding this,
Their lips drew near, and clung into a kiss.

Had Don Juan forgotten his most recent conquest?

And should he have forgotten her so soon?
I can't say but it seems to me most truly
Perplexing question; but, no doubt, the moon
Does these things for us, and whenever newly
Strong palpitation rises, 'tis her boon,
Else how the devil is it that fresh features
Have such a charm for us poor human creatures?

Thus England's "moon-struck bards" remade the moon in their own image—lonely, wandering, and seductive. Or as Byron, describing many a "tender moonlit situation" noted:

But lover, poet, or astronomer,
Shepherd, or swain, whoever may behold,
Feel some abstraction when they gaze on her:
Great thoughts we catch from thence (beside a cold
Sometimes, unless my feelings rather err);
Deep secrets to her rolling light are told;
The ocean's tides and mortals' brains she sways,
And also hearts, if there be truth in lays.

Keats died of tuberculosis in 1821. Shelley drowned the following year. Byron soon died of fever and Blake of old age. Scientists buried Goethe's theory of color, denouncing it as "a striking example of the perversion of the human faculties." As the Romantic era faded, Haydn's *Creation* was rarely performed, and would soon be mocked as "a third-rate oratorio." Turner's most primal paintings could not be found in any museum. The lone legacy of light's Romantic interlude was the golden, soft, tender, lovely, gracious, splendid, beguiling moon.

In 1832, Beethoven's somber Arbor Sonata was renamed the Moonlight Sonata. Byron's *Don Juan*, wooing a new generation, made the moon a fixture on lovers' lanes. Younger writers, most of them French, took up where Byron left off, setting their romances beneath full moons. Popular songs have since done the same. The Romantics had given the moon a fresh purpose. No longer a symbol of fertility but the call to it, moonlight swayed neither fate nor minds, but hearts. In 1832, on his deathbed, Goethe uttered his last request—to please open the shutters so that he might see "more light."

Part Two

We must learn to think more subtly than in the past.

—NIELS BOHR

CHAPTER 10

Undulations: Particle vs. Wave

> I must own I am much in the dark about light. I am not satisfied with the doctrine that supposes particles or matter called light are continually driven off from the sun's surface with a swiftness so prodigious.
>
> —BENJAMIN FRANKLIN

Throughout fifty centuries of civilization, light remained untamed and largely unknown. Those who saw it as God or the face of God had to rely on faith and those who studied light were, as Newton saw himself, children playing onshore "whilst the great ocean of truth lay all undiscovered before me." A cabinet of tools had tinkered with light. Starting with sticks planted in the ground, the curious had moved on to mirrors of polished silver, lenses of gritty glass, Ptolemy's coin in a cup, Alhacen's barley grains, Galileo's telescope, and Newton's prisms. Light had intrigued students in togas, in turbans, in cloak and cowl. It had engaged a university of disciplines—mythology and religion, philosophy and astronomy, architecture and painting, physics and metaphysics, poetry and prose, geometry and trigonometry . . . Yet at the start of the nineteenth century, here is all that was known about light:

- It traveled in straight lines, except when it didn't (diffraction).
- Its angles of deflection could be calculated.
- Its brightness, like gravity, diminished according to the inverse square law.

- It could be broken into colors.
- It traveled 144,000 (or 196,000) miles per second through air, and still faster in water. (Or perhaps slower.)
- It flowed through a luminiferous ether no one had proved to exist.
- It was composed of particles. (Or waves?)

Then, beginning in 1801, a new generation of students began probing light. While Goethe sought the soul of color and Byron romanticized the moon, fresh experiments and revived skepticism helped light come of age. And in far less time than the Greeks spent arguing over eidola, light evolved from nineteenth-century curiosity into twentieth-century tool. The discoveries sometimes collided and canceled each other in confusion. More often, however, they were in sync, adding force to force like waves overlapping.

Modern students of light were as eccentric as their forbears. One was a brilliant linguist who helped decipher the Rosetta stone. Another, when not refining quantum mechanics, spent his spare time banging on a conga drum. One was a self-described "calculating machine" who also pumped out rhyming verses of Victorian grace. And the most famous became the embodiment of eccentric genius, using light to unsettle time, space, matter, and motion. But he refused to wear socks. All these modern students shared the spirit of the physicist Albert Michelson, the first American scientist to win the Nobel Prize. Asked why he studied light, Michelson replied, "Because it's so much fun."

Yet the deeper enigmas surrendered not just to genius. The struggle to think more subtly about light involved jealousy, feuds, contests, and a crystal found on the coast of Iceland.

Legend has it that the Vikings used a "sunstone" to steer their ships on cloudy days. A sailor held this crystal to the sky, turned it ninety degrees, and used some directional property of daylight to find his way. Perhaps. But the rest of Europe did not discover Iceland spar until the 1660s, when a traveler brought a few translucent chunks back to Copenhagen. Many puzzled over the little crystal. When placed over a printed page, it doubled the letters, making ghostly words hover inside it. Someone noticed that Iceland spar split candlelight, sending two beams in different directions. Isaac Newton soon got word. In *Opticks* he noted how "that strange substance, Island Crystal" divided beams. Shining light through two crystals, Newton noticed another oddity. If their broad faces were parallel, the second spar also split the first one's beams. But turn one spar ninety degrees and it split one beam but not the other. Unable to explain, Newton fell back on hypothesis. It seemed that "Every Ray of Light has therefore two opposite Sides."

What Newton could not explain with particles, Christiaan Huygens tried to calculate with his wave theory. Huygens's *Traité de la lumière* included an entire chapter on Iceland spar. "Amongst transparent bodies," the Dutch astronomer wrote, "this one alone does not follow the ordinary rules with respect to rays of light." The bulk of Huygens's treatise is thick with equations, coldly clinical, yet he found the spar's tricks "marvelous." Using the Snell-Descartes law of refraction, he calculated the angles of waves passing through the spar and splitting into "the ordinary" and "the extraordinary" refraction. Huygens updated Euclid's origami with diagrams that resembled windmills, with waves circling squares and triangles. These showed how a single crystal could have two different refraction indexes, cleaving a single beam into two. Yet when pondering why a second spar split the "ordinary" but not the "extraordinary" ray, Huygens was stumped. "It seems as though, as it passes the upper plate, the ordinary ray has lost something that is necessary to bring the matter into motion which is needed for the irregular refraction but to say how that operates—up until now I have discovered nothing that satisfies me."

Other investigators, learning of the spar from Newton or Huygens, made their own studies. A British physicist tested Iceland spar in a wooden contraption with angled mirrors and prisms. A French student dropped his spar and, intrigued when it split into perfect cubes, went on to pioneer the field of crystallography. Another explained the spar's curiosities as optical illusions. (And I bought a chunk of Iceland spar on the Web and aimed my trusty laser pointer at it. One red beam hit the far wall and a second, fuzzy spot shone a foot to the left.) All came away defeated by the curious little crystal. Newton's particles could not explain its effects, nor could Huygens's waves.

By 1800, the study of light had reached a standstill. Nearly a century had passed since Newton's *Opticks*, and in that century the only breakthrough with light came not in physics but in botany. In 1779, a Dutch physician, Jan Ingenhousz, immersed green leaves in sunlit water. The leaves gave off bubbles, but the bubbles stopped when set in shade. Ingenhousz praised each plant's "great power of purifying the common air in sunshine." This was light's greatest blessing—oxygen created by photosynthesis. Physicists, however, were still stuck on Newton.

Common wisdom assumed that light had been mastered. If there were anything more to be learned, surely Newton would have found it. Newton's authority reigned even as the Enlightenment spiraled into protest and revolution. By the start of the 1800s, however, those just beginning their scientific careers were sick of Newton worship. Perhaps with the latest instruments, the latest math, there might be more to learn about light. Its very essence might even be known.

In discussing his own intellect, Thomas Young deferred to the praise of others. Fellow students at Cambridge in the 1790s called him "Phenomenon Young." Later this quiet British gentleman would be recognized as "one of the most acute men who ever lived." But Young liked to tell people, with no pun intended, that he "may be said to have been born old, and to have died young." Able to read as a toddler, he mastered Latin and Greek by the age of six and was soon reading Horace and Virgil in the original. He went on to learn a dozen languages. At fifteen, along with devouring Euripides, Sophocles, Euclid, Homer, and a history of France, he plowed through the works of Newton. Reading *Opticks*, the teenager noticed "one or two difficulties in the Newton system." At universities in Edinburgh and Göttingen, Young set aside light to study sound. His thesis on sound waves firmly grounded him in the complex geometry of wave theory. Drawn to medicine, Young studied the human eye and soon devised his own theory of color vision—that the retina has just three types of color receptors, for red, green, and blue. All the myriad colors we see are combinations of these three. Opticians did not confirm this for 150 years. With waves and vision behind him, Young naturally turned to light, taking up where Newton had left off: diffraction.

In Bologna in 1661, the discoverer of diffraction, Father Francesco Grimaldi, had puzzled over another of light's quirks. Beams shining through adjacent holes made bright overlapping circles, yet each overlap contained dark vertical lines. Newton likewise saw black gaps between his colored rings. His only explanation was that light passed in "fits"—"fits of easy Reflexion" or "fits of easy transmission." With all due respect to Newton, Young suspected a different cause.

The nineteenth century was barely begun when, on November 12, 1801, twenty-six-year-old Thomas Young, trained in medicine but not in optics, stood before the leading scientists of London to inform them that Isaac Newton was wrong. Reading his treatise "On the Theory of Light and

Colours," Young announced that light did not consist of particles. For how could it be that particles of light, whether emitted by "the white heart of a wind furnace, or the intense heat of the sun itself . . . are always propelled with one uniform velocity?" And why, when light struck water, did some particles bounce off while others passed through? Young also rejected Newton's conclusions about Iceland spar, preferring Huygens's wave calculations. Young then turned to Newton's rings. Studying their gaps "has converted that prepossession which I before entertained for the undulatory system of light." No one knows whether this challenge caused Newton's disciples to shout or merely clear their throats, yet anyone doubting Isaac Newton had more explaining to do.

Young continued, comparing light to water. "Suppose a number of equal waves of water to move upon the surface of a stagnant lake . . ." Young described what all present had seen—waves overlapping, combining strength, or sometimes, when crest met trough, canceling each other, leaving still waters. "Now I maintain that similar effects take place wherever two portions of light are thus mixed, and this I call the general law of the interference of light."

When Young finished, the meeting continued with the usual propriety. Young's reputation—he had been admitted to the Royal Society at age twenty-one—commanded respect, but further proof was required, and Young soon offered it. In 1802 he invented the first new optical tool since the telescope, the ripple tank. Here was one of those elegant devices that stirred wonder—especially wonder at why no one had thought of it before. The ripple tank—an aquarium propped on a rack, with a candle beneath, a mirror above, and a screen in front—turns wave motion into patterns easily projected. Jiggle a bar in the water and waves flow, their curves shimmering across the screen. Today, Young's ripple tank is a staple in physics classrooms and is easy to make. (Or download Ripple Tank, a free iPad app that displays wave patterns in a variety of cool colors!) The ripple tank confirmed Young's suspicions. On his screen, he saw spreading waves collide and "interfere," leaving cross-hatchings where all was still. But waves in a tank did not prove light itself to be waves. Young needed his own *experimentum crucis.* What would happen, he wondered, if he beamed light through two slits side by side?

What is now known as "Young's experiment" has been called "perhaps the single most influential experiment in modern physics." If light was waves that crisscrossed and collided, then this "interference" should display itself in predictable patterns. Where two waves were in sync, their crests paralleling each other, boosting each other, the light should be brighter. Where the crest of one wave met the trough of another, they should cancel each other, creating darkness, or at least dimness. And when he sent parallel beams of a single color through adjacent slits, Young saw on his wall the pattern he expected—a bright central stripe flanked by zebralike dark and bright lines. In his proper prose, Young concluded, "We may be allowed to infer that homogeneous light at certain equal distances in the direction of its motion is possessed of opposite qualities capable of neutralizing or destroying each other, and of extinguishing the light where they happened to be united."

Light on light creating darkness? No particles, no fits, no faith in Newton could explain this. To support his theory, Young made the first measurement of light waves. The math was complex, the logic simple. If light is a wave, then each undulation must have a precise length, the distance between crests. For two waves to interfere destructively, Young reasoned, one crest must strike the other's trough, like a riptide receding from shore to stop an incoming wave cold. Measured from the central bright stripe, dark lines should appear at distances proportional to a half-wavelength of light or any odd numbered multiple of that. Young measured his stripes and their distance from his candle, then shifted the distance between his slits and measured again. Applying the geometry of similar triangles, he arrived at an answer no less astonishing than light's speed. Sound waves, Young knew, could be meters long, but light was its own strange entity. The length of a red wave proved to be unfathomably short—just 0.00000065 meters. Violet was shorter still—0.00000044 meters. Young's measurements, though made with instruments that today seem Stone Age, were within nanometers of modern calculations. Light waves now had lengths, yet they still fell within Newton's shadow.

Young published his discoveries in 1807. Defining his "undulatory system," he sketched a diagram of intersecting waves like those made by stones tossed in a pond. All could see the Xs where the waves crossed, making no disturbance, no light, but Newton's faithful chose not to see. Young's theory was blasted as "one of the most incomprehensible suppositions that we remember to have met with in the history of human hypotheses." Young fired off a response, but no newspaper would print it. He self-published a pamphlet. "Much as I venerate the name of Newton," he wrote, "I am not therefore obliged to believe that he was infallible." The pamphlet sold one copy. Young then fell back on Newton's own defense. "Let him make the experiment," he said of a critic, "and then deny the result if he can."

Young soon turned to other pursuits. Having earned his medical degree, he set up a practice near the English Channel and resumed his study of languages. In 1814 he focused on a remarkable stone, etched with three separate scripts, which British soldiers had brought home from Egypt. Young had no trouble reading the ancient Greek, yet neither he nor anyone else could decipher Egyptian hieroglyphics. Young became obsessed with the stone's third language, demotic, and after a year of study, had a clue. Archaeologists already suspected that two languages carved on the stone conveyed the same text, but Young saw that demotic bore a "striking resemblance" to the hieroglyphics, and that the Egyptian symbols were phonetic, each standing for a spoken sound. Using Young's insights, Jean-François Champollion soon deciphered the inscriptions on the Rosetta stone. Meanwhile, news of Young's interfering light waves spread across the English Channel. There it challenged not just Newton but the entire history of optics.

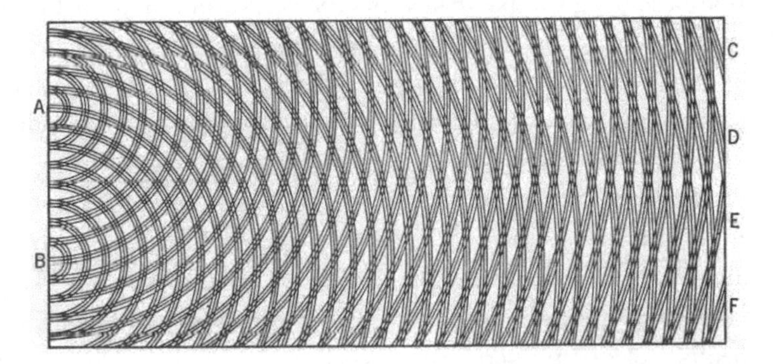

Ever since Empedocles' lantern beamed "its tireless rays," the ray, not the wave, had been the fountain of light. Euclid's *Optics* began "The ray issues from the eye . . ." Buddhist scripture spoke of rays like jewels intermingled, and Augustine saw God as a "ray of certain intelligible Light." Al-Kindi's rays radiated and Alhacen's were likened to balls thrown at wood. If light were to be suddenly seen as waves, it would have to do more than make stripes. The very nature of light was at stake. Particles are matter; waves are motion. Everyone knew sound was a wave, but particle theorists, soon known as "emissionists," argued that sound behaved differently than light. You could hear a shout from around a corner, but you could not see the shouter. And sound depended on air. A bell in a glass chamber fell silent when the air was pumped out, yet it could still be seen. Newton had seen diffracted light behave "with motions like that of an Eel," yet only particles, he concluded, could travel in straight lines. "Sounds are propagated as readily through crooked Pipes as through streight ones," Newton wrote, "but Light is never shown to follow Crooked Passages nor to bend into the Shadow."

Still, what about that little crystal from Iceland?

Once England rejected Thomas Young's undulations, light's torch was passed to France. Having survived its revolution, having since conquered and lost half of Europe, France would spend the next century more enthralled by light than any other nation since ancient Greece. From Paris and its environs would come the first light shows, the first photography, the first modern lighthouse, the best measurement of light's speed, the freshest use of light on canvas, and the first motion pictures. The spark was struck in a single school.

Back in the summer of 1794, as France was besieged by mobs and mass executions, the Ecole Centrale des Travaux Publics quietly opened in the Latin Quarter of Paris. Its goal, like that of the French Revolution, was to engineer the future. Within a year the school was renamed the Ecole Polytechnique, and to its gates came the top technical minds in France. Proud, patriotic, and exacting, these professors forged a rigid curriculum based on calculus, civil engineering, chemistry, physics, and optics. Optics was poorly taught at the Ecole, its professor "a professor only in name," as one student said, but its place in the canon made all students—future chemists, physicists, engineers, mathematicians, biologists—curious about light. Ecole professors believed Newton's particle theory as faithfully as they trusted his laws of gravity and motion. Even when Napoleon turned the Ecole Polytechnique into a military academy, it continued to graduate brilliant, feisty men willing to risk careers and friendships to defend Newton. The names of this rising generation are written in the history of science and on the Eiffel Tower, seventy-two surnames stamped in iron just above the massive arches:

AMPERE FOUCAULT FIZEAU FRESNEL ARAGO

A few were savants sequestered in labs and classrooms, but many were also diplomats, Parisian sophisticates, and adventurers crossing deserts and soaring in balloons. Napoleon, recognizing them as France's best and brightest, commandeered forty-two Ecole students to accompany his conquest of Egypt. Graduates of the Ecole Polytechnique could expect no sinecures, no tenure, only teaching, travel, and endless arguments over scientific truths held as close to their hearts as they held France itself. Yet the Ecole soon developed a generation gap. Professors who had come of age under Newton's sovereignty feared students eager to challenge him. By 1807, the challenges were mounting. First came arguments over Iceland spar, then word of Thomas Young's interference theory. Both might be explained or explained away, but further challenges would be harder to refute. The only way to prove the upstarts wrong, the elders decided, was a competition.

In December 1807, the Académie des Sciences proposed its latest competition. Entrants were to use Christiaan Huygens's wave theory to explain the split rays of Iceland spar. Just a handful of young scientists tried. The one who succeeded had survived battles far more brutal than Particle vs. Wave.

Etienne-Louis Malus is hardly a household name, yet hardly a household is without a device his discovery made possible. Malus had turned to light in 1798 as an escape from the horrors he saw when yanked out of the Ecole Polytechnique and sent into Egypt with Napoleon. Malus cut a dashing figure, with dark, curly hair, muttonchop whiskers, and stiff epaulets, yet he was an indifferent soldier. After contracting bubonic plague, he was shipped to the pesthouse in Jaffa and then back to Cairo. There he wrote in his journal of "the tumult of carnage . . . the smell of blood, the groans of the wounded, the cries of the conquerors . . ." He watched friends succumb to the plague and saw their discarded bodies ravaged by wild animals. Somehow he survived and was sent back into battle. Seeking hope amid misery, he turned to light. Late into the night, on the sands of the Nile Delta, his tent glowed as he manipulated candles and mirrors, calculated angles, and kept his spirit alive even as his body struggled with dysentery and other diseases. Later, back home in Paris, Malus continued his studies. Holed up in his room near the Luxembourg Gardens, he became enthralled by Iceland crystals. Beaming light through chunks of spar, he noticed yet another startling trait. A spar bisected all beams *except* those that struck its surface at a precise angle. At 52 degrees, 54 minutes, the light shot straight through.

When Malus learned of the Académie competition, he spent a year obsessed with light. "I live here like a hermit," he wrote. "I pass whole days without speaking a word." In a room littered with candles and crystals, he etched a sheet of copper with a scale in millimeters. He then set a spar on the scale and measured each angle of each double image. Blending the law of refraction with some advanced algebra, he piled equation on equation. Light had been calculated since the days of Euclid, but never with such precision. In December 1808, Malus submitted his entry. Académie elders, though loyal to Newton, could not refute Malus's math. He won the competition. Wave theory could explain Icelandic spar, but the embattled soldier was not done with his crystals.

One bright fall afternoon, Malus sat, silent and sullen, in his room. Spotting sunlight glinting off the windows of the Luxembourg Palace, he lifted a piece of spar to his eyes. Like Goethe seeing the world through a prism, Malus was stunned. He expected the sun's reflection to be doubly refracted, yet through the crystal he saw a single image. That evening, still peering through spars, Malus stared at a candle flame reflected in water. Stooping and standing to adjust the angle of incidence, he looked, checked, looked again. At the spot where candlelight struck water at 36 degrees, the double refraction turned to single. Light reflected off glass or water, Malus concluded, somehow *changes*.

Consulting trigonometry tables, calculating the sine and cosine of angles, Malus concluded that, as Newton had suspected, light has "sides." Light must be asymmetrical. For this phenomenon Malus then coined a term we still use—"polarized light." When light's "sides" are aligned, they pass through some transparent surfaces but are filtered by others. To test his theory, Malus built a device with rotating mirrors, one above the other along parallel axes. Now he could bounce light at any angle. With a little more math ("the cosine squared of the planar angle"), a bit more insight (if you rotate a spar 90 degrees the "ordinary" ray behaves like the "extraordinary," and vice versa), Malus proved the principle of polarization. Light was not like Descartes's tennis balls but more like a football (American) that wobbled end over end or, under certain conditions, flew in a perfect spiral. Light passing through Iceland spar or reflected at precise angles became polarized, its "sides" aligned.

Polarization was announced by the Académie in 1811. Etienne-Louis Malus died the following year, at age thirty-seven, of complications from diseases contracted in Egypt. His name is known primarily to scientists, but every decent pair of sunglasses filters polarized glare, and every liquid crystal display in a flat screen TV or laptop uses polarized light to brighten and darken its pixels. (Don sunglasses and look at your laptop. Now tilt your head. If you don't see the screen darken, those are cheap sunglasses.) Physicists would soon extend Malus's discovery, exploring various kinds of polarized light—circular and elliptical—paving the way for the light of the twenty-first century. But these advances depended on wave theory, and that paradigm was still shifting. Mounting evidence was beginning to shake even the assurance of the French. Resistance to Newton was rising. Académie patriarchs kept a careful eye on the upstarts.

One such upstart was François Arago. Born in southern France, Arago was a Parisian outsider whose Spanish surname led some to question whether he was "really French." Arago had enrolled in the Ecole Polytechnique in 1803 to study astronomy. His first work after graduation was proving that sunlight traveled at the same speed as light from other stars. In 1806, the tall, dashing astronomer was sent to Spain to continue a long-running project, measuring the earth's meridian of longitude. Beaming light from distant towers, Arago triangulated the results into precise distances. But when Napoleon's army invaded Spain in 1807, Frenchmen shining lights from towers were suspected of espionage, and Arago was jailed on the Isle of Ibiza. He escaped twice, only to be recaptured. Finally freed, he was sailing toward Marseilles when a strong wind blew him to North Africa. Arago hired a guide to lead him, disguised in a burnoose, across the desert. When he arrived in Paris, his story made him a celebrity. Handsome, charming, and unquestionably French—he would later serve as prime minister—Arago was admitted to the Académie des Sciences. He continued to study light but soon discovered that during his ordeal abroad, the Académie had become a battlefield.

Members of England's Royal Society often behaved like boys, but France's counterpart was a roiling stew of ego and genius. Meeting at the Institut de France, overlooking the Seine, Académie members wore their frock coats with starched disdain. In an expansive room graced by statues and parquet floors, members presented papers, awarded patents, and passed judgment on what the public should know and believe. But when it came to something as important as light, decorum surrendered to backbiting and outright theft.

In 1812, beaming light through thin sheets of mica, François Arago discovered chromatic polarization. Just as sunlight can be polarized, so can the rainbow and all refracted colors. Arago announced his discovery to the Académie, but before he could publish it, his partner from the expedition in Spain, Jean-Baptiste Biot, protested. Like Robert Hooke usurping Newton, Biot insisted that *he* had already discovered *polarisation colorée.* An Académie investigation sided with Arago, but by then Biot had published his version and his story stuck. Arago fumed, Biot went on a PR blitz, and the two men stopped speaking to each other, though they often shouted during meetings. Though a Newtonian, Arago was suddenly open to any theory of light that might deflate his rival. His openness was rewarded in December 1814 when, at a posh state dinner, he first heard the name Fresnel.

Seated at Arago's table, a stranger spoke of his nephew, a sickly sort yet quite promising. The frail young man was a civil engineer building roads and bridges in a remote part of France. For the past year he had spent whole evenings studying light. He had recently sent a paper to M. Ampère, but M. Ampère had not responded. Would M. Arago take a look?

Only six months earlier, Augustin-Jean Fresnel (pronounced "Frenel") had written to his brother confessing his confusion about light. Professors at the Ecole Polytechnique had insisted that light was made of particles, yet Fresnel was beginning to doubt. He had heard of polarized light, but admitted, "I've about broken my head over it, but I can't divine what this is." When his brother sent him optics textbooks, Fresnel began drifting toward renegade theories. "I tell you I am strongly tempted to believe in the vibrations of a particular fluid for the transmission of light and heat," he wrote his brother. "One would explain the uniformity of the speed of light as one explains that of sound; and . . . why the sun has for so long shined upon us without diminishing its volume, etc."

Although doubting Newton, this French civil engineer was much like him. Newton was known as "fearful, cautious, and suspicious." Fresnel's colleagues considered him *un homme froid*, a cold man. Both Fresnel and Newton were plagued by bad health. Newton imagined most of his maladies, but Fresnel, short and fragile with aquiline features and fugitive eyes, constantly struggled with coughs and fevers. The two men also shared an intuitive sense of how to isolate beams in a dark chamber, making light behave just as they expected. And like Newton, Fresnel had what he called "a taste for exactitude" that sent his equations sprawling across page after page.

Between 1814 and 1819, the First Light of Haydn's *Creation* blared in symphony halls throughout Europe. John Keats wrote *Endymion* ("What is there in thee, Moon . . .") and J. M. W. Turner absorbed the radiance of Venice. Meanwhile in a small room a dozen miles from the English Channel, Augustin-Jean Fresnel sat alone, often shivering, calculating—to the third, fourth, and fifth decimal place—*exactly* how light traveled.

Fresnel began with the familiar setup—a pencil of sunlight fed through a pinhole. Sunlight through pinholes had been steady enough for Ptolemy, Alhacen, and Newton, but the sun moved too quickly for Fresnel's level of precision. His micrometer, made by a local blacksmith, measured to 0.01 millimeter, but only if he could somehow slow the passing light. A lens did not help, but when Fresnel dabbed the pinhole with honey—his mother kept bees—the sunlight refracted slightly and lingered. To examine diffraction, he aimed light at a taut thread. The thread cast lines on his wall, dark in the middle, softer above and below—a sandwich of shadows. Fresnel took out his micrometer. "Using a lens of 2 mm. focus and light which is practically homogeneous, I have been able to follow these fringes very close to their origin . . . The interval which separates this band from the edge of the shadow I have measured on the micrometer and find it to be less than 0.015 mm." Only when Fresnel came to Paris did he learn of Thomas Young's paper on interference. Unable to read English, Fresnel did not look at it. Yet back in his room, Fresnel saw patterns of interference, and another of Young's discoveries. Blocking the light above his thread, he expected only the upper layers to disappear. Yet *all* the fringes, above and below, vanished. Clearly, diffraction required the passing of waves both over and under an edge. Particles would not behave this way. Fresnel sent his conclusions to Paris. M. Ampère did not read them but M. Arago was intrigued. Here was a fellow upstart, from beyond the sniveling Académie, who might help him challenge his smug rivals. Arago set out "as much as the state of the sky permitted" to verify Fresnel's discoveries. Soon the celebrated astronomer wrote to the unknown engineer, saying his results might "serve to prove the truth of the wave theory."

Arago told colleagues about Fresnel and presented his paper to the Académie in March 1816. That summer Fresnel returned to Paris, met with Arago, and the two finely tuned Frenchmen talked and talked about light. Fresnel set up a lab in Paris, but 1816 was the "year without a summer." A volcanic eruption in Indonesia had coated the atmosphere with ash, leading to year-round freezes, famine, and a drizzlier-than-usual summer in Paris. Frustrated, Fresnel went home and resumed his engineering duties. A few months later he heard again from Arago, who had just visited Thomas Young in England. Arago recounted how, when he had been explaining Fresnel's work to Young, the Brit had kept a polite silence, but Mrs. Young stood and left the room. Moments later, she returned with her husband's pamphlet, open to the page detailing his similar experiments on interference. The implication was clear. Reading it, Fresnel considered giving up his studies. "I have decided to remain a modest engineer of bridges and roads," he wrote his brother. "I now see it's a stupid plan troubling oneself to acquire a small bit of glory, that they'll always quarrel with you about it." But by the following spring, he was back in his room, unable to rest until he solved the essential problem posed by waves.

For wave theory to be plausible, the math had to work. Passing through water, waves had to refract according to the Snell-Descartes law. Bouncing off mirrors, they had to reflect at the angle of incidence. Christiaan Huygens had calculated wave fronts marching like soldiers, advancing in line through open space. Huygens's waves moved predictably, mathematically, according to the few laws of light known before 1700. But could a wave, like water, swirl around an object and continue in a straight line? Everyone knew that water striking a barrier surged backward. Could anyone imagine light forming such eddies? If wave theory itself were to advance, these patterns had to be proved on paper. Huddled in his mother's house near the English Channel, Fresnel set out to try.

While Fresnel refined his measurements, Newton's disciples in the Académie marshaled their defenses. The aging scientists—POISSON LAPLACE LAGRANGE—knew Newton to be right about everything—about gravity, about motion, about the calculus, the prism, and the rainbow. Now they were asked to believe that an unknown civil engineer, not yet thirty, who held no teaching nor research position, had a better grasp of light than their "philosophic sun." Such nonsense might go on indefinitely unless put to another competition

On March 17, 1817, the Académie des Sciences announced the challenge. Entrants were "to determine by precise experiments all the effects of diffraction of luminous rays both direct and reflected, when they pass separately or simultaneously, near the extremities of one or more bodies . . ." In other words, entrants had to calculate, using the most refined math, how light traveled around an obstruction. As Fresnel read the rules by lamplight, one word kept leaping out. *Rayons.* "Luminous *rays* . . ." "Light from which the *rays* emanate . . ." "The motions of the *rays* . . ." Clearly, the Académie was not interested in hearing about waves, and Fresnel hesitated to enter. François Arago pressured his friend, and M. Ampère, whose name would later become a measure of electricity, agreed. "He will easily win the prize," Ampère wrote to Fresnel's uncle, who urged his nephew to join the "hard battle" led by "General Arago." Contestants were given eighteen months. Fresnel took a leave of absence from his engineering post.

He began by attacking particle theory. It could not account for all of light's motions because particles themselves had limited motion. Arrows, tennis balls, or *rayons* bounced and bent as predicted, but neither would diffract around a sharp edge nor interfere with other light to cause dark stripes. "One can add a motion of rotation to that of transmission; but that's all," Fresnel wrote. Those who clung to particles, he believed, were succumbing to a longing as old as humanity itself—for simplicity. But nature, he recognized, "does not dread difficulties of analysis." The difficulties came quickly.

Huygens's wave theory had not used the calculus. While working on his theory in the 1670s, Huygens had tutored young Gottfried Leibniz in math. Leibniz, who later devised his own version of the calculus, often wrote

Huygens to tell him of its possibilities—the mastering of infinite regressions and instantaneous velocities, the graphing of nature's subtlest forces. Yet Huygens, schooled in the analytical geometry of Descartes, whom his father had known personally, never fully accepted the calculus as a new branch of math. He published his *Traité de la lumière* three years after Newton's *Principia* but chose not to update it with calculus. Now, more than a century later, Fresnel did, using Newton to disprove Newton.

Fresnel's entry to the Académie contest tapped his earlier work, yet he labored for months to recalibrate the math. The key was the calculus tool known as the integral. An integral, denoted in an equation by an elegant S ($\int$), measures curves and the areas they sweep out. Integrals calculate a missile's trajectory, the graceful shapes of seashells, and, as Fresnel saw, the motion of a wave. Applying calculus to light waves, Fresnel crafted what are now called the "Fresnel integrals." To the novice they look like modern-day hieroglyphics, a lattice of the highest math that sets the mind spinning. To stare at Fresnel's equations is to feel yourself drawn toward the infinite complexity of the universe. Symbols and signs are stacked on each other like layers of a cake. Then come the Greek letters, not just the familiar pi (π) but lambda (λ—wave length), sigma (Σ—defining a summation of numbers), and kappa (κ—curvature). All are crowded inside parentheses, crowned by exponents, huddled in brackets within brackets. The whole is far greater than the sum of the parts, and if you stare long enough, Fresnel's concerto of calculus is both hypnotic and inspiring. To think that *this* is how light behaves, to know that a single man opened this door is to grasp light's complexity and marvel at human discovery.

And yes, light does swirl and eddy like water. Building on Huygens, Fresnel calculated how waves create secondary fronts. Imagine hearing, through an open door, a shout from the next room. The sound seems to originate from the door itself—a new wavelet. Yet part of the original sound wave stays in the first room, flowing back on itself. Or, as Fresnel put it, if "the molecules were in their equilibrium positions [undisplaced] and received at that instant only the speeds that push them forward, a wave toward the rear would also result . . . These two motions therefore counter one another in the retrograde waves."

On April 20, 1818, the battle of Particle vs. Wave approached its barricades. Using a Latin epigram *Natura simplex et fecunda*, "Nature simple and fertile," Fresnel submitted his contest entry in a sealed envelope. Three months later, an anonymous entry showed up. These were the only two attempts to disprove Newton. A panel of judges convened. Three were staunch "emissionists." Another, a young chemist, was undecided. The fifth member, the chairman, was François Arago. The committee deliberated throughout summer, fall, and into the new year. To make Fresnel's entry more palatable, Arago had changed each mention of "waves" to "elementary rays." Still the discussion continued. One emissionist, the esteemed mathematician Simeon Poisson, did his own calculations. If light beamed at a disc behaved as Fresnel asserted, Poisson argued, it would leave a bright spot in the dead center of its shadow. *C'est absurd!* Emissionists thought they had trumped Fresnel, but Arago molded a freckle-sized disc and aimed a beam at it. On the wall behind, there at the center of its shadow—the dead center—was a perfect pinpoint of light. Poisson refused to budge—he believed in the particle theory for the rest of his life—but the committee was convinced. Eleven months after submitting his entry, Fresnel was declared the winner. He cared little for the honor, later telling Thomas Young that acclaim did not compare with the thrill of discovery. But from the moment Fresnel won the Académie contest, particle theory was in retreat.

Over the next several years, despite a hacking cough, Fresnel intensified his studies. He focused first on Iceland spar, adding to its enigmas. When he sent light through the crystal, its beams polarized as expected. Yet while some interfered with each other, making the telltale stripes, beams polarized at right angles did not. Relying as much on intuition as on math, Fresnel proposed a property of light that Thomas Young had rejected. Wave theorists assumed that light waves advance along a parallel front like an open hand sweeping salt off a counter. Sound was known to be such a wave. Fresnel suggested an alternative. Light waves might be transverse, whipping up and down like a jiggled rope tied to a doorknob. Arago refused to believe this, but Fresnel used calculus to prove it.

Just seven years after breaking his head over polarization, this delicate, sickly civil engineer knew more about light than anyone else in the world. The name Fresnel now graces not only the Eiffel Tower but also several optical concepts, including Fresnel diffraction, the Fresnel spot, and Fresnel lenses found in automobile headlights, cameras, projectors, and solar collectors. The fantastic Fresnel integrals are still used to calculate waves ranging from the curves of roller coasters to the lapping of ocean waves simulated in video games. But Fresnel's most enduring legacy was his outdueling Isaac Newton to prove that light flowed in waves. The flow would last for the rest of the century, a fabulous century when light became a master showman.

CHAPTER II

Lumière:
France's Dazzling Century

The urchin of present-day Paris . . . is at once a national emblem and a disease. A disease that must be cured. How? By light.

—VICTOR HUGO, *LES MISÉRABLES*

Early in the nineteenth century, sailors on tall ships entering the North Atlantic knew they were bound for a dark and dangerous place—France. The revolution was long past, but stories lingered—of the guillotine, of riots and rage, of Napoleon, that scourge of all Europe. Paris, owing to its role in the Enlightenment, was coming to be called the City of Light, yet Paris itself was a scourge, a labyrinth of disease and decay, its air reeking of emptied chamber pots, its river yellowed by raw sewage. City of Light? Not in the warrens where most lived, certainly not in winter, when, as in *Les Misérables*, "the sky had become a grating, the day a cellar, and the sun a poor man at the door."

The dangers of France were not confined to Paris. The nation's rocky coastline had few lighthouses, each so dim that a ship was almost upon it before seeing its beam. Each year nearly two hundred ships ran aground off French shores. Still, sailors drifted toward France's western flank. There the river Gironde led to the city of Bordeaux, whose wines the world demanded. As the coast loomed, the search began for the lighthouse—*le phare*—at the river's mouth.

The Cordouan lighthouse was one of Europe's oldest and tallest. Perched on an outcropping a day's sail north of Spain, Cordouan was "the Versailles of the sea." In regal French style, the ten-story tower featured a first-floor bedroom readied for the king, a second-floor chapel, and abundant statues throughout. The light, however, was no stronger than when it first shone back in the age of Descartes. Then it had been a bonfire, but in recent years, keepers had experimented with lamps and mirrors whose puny light had sailors begging France to make this *phare* worthy of its namesake, Alexandria's Pharos. Napoleon established a lighthouse commission, but it rarely met. Complaints went unheeded. The Cordouan light remained a flickering candle. Then, in 1820, the world's leading expert on light began studying the problem.

The problem was reflection, Augustin-Jean Fresnel concluded. Even the largest parabolic mirror placed behind the strongest lamp reflected less than half the ambient light. Applying his elegant integrals and his "taste for exactitude," Fresnel calculated how to capture light by refraction rather than reflection. Within months, glass factories in Paris were turning out prisms three inches thick and twenty inches long. Following Fresnel's design of "lenses by steps," glassmakers assembled polygonal plates two feet across, each ribbed with crystalline circles that focused light from all directions. By April 1821, Fresnel's lens, a twelve-foot tall glass beehive, was ready for testing. Placed atop the Royal Observatory south of Paris's Luxembourg Gardens, the lens shot a beam over the Seine, past Abbot Suger's Saint-Denis, and beyond the hills of Montmartre. Within a year, several hundred pounds of prisms and glass plates were shipped around the cliffs of Brittany and landed at the Cordouan light. Fresnel, though sometimes coughing up blood, spent an entire spring on the wind-whipped outcropping at the river's entrance.

On July 25, 1823, the dark and dangerous coast of France shone with the brightest light yet made by man. The sweeping white spear, easily visible to the horizon, was also seen by sailors atop mastheads more than thirty miles out to sea. Two years later, British sailors saw a Fresnel light beaming over the English Channel "like a star of the first magnitude." Fresnel soon succumbed to tuberculosis, dying at age thirty-nine, but his light lit coasts from Sweden to Spain to New Jersey. And having shown the way for sailors, France became the world's keeper of the light. As if following the old saying *rayonnement de la France*, ("the shining of France"), the French captured, reflected, projected, shadowed, shifted, painted, clocked, froze, and framed all manner of light. The same fascination that had gripped professors and students at the Ecole Polytechnique spread to the public at large, turning light into a spectacle, a time machine, and ultimately a gift to the modern world.

* * *

Greeting the sunrise in Varanasi (Benares), India. One of many prayers to the dawn in the Hindu Rig Veda sings praise: "We have beheld the brightness of her shining; it spreads and drives away the darksome monster." (Getty Images)

The first Gothic cathedral, Saint-Denis in Paris, used architectural innovations to bring the light of heaven indoors. Opened in 1144, Saint-Denis inspired hundreds of imitators, defining Gothic light for the Middle Ages. (Getty Images)

Paradiso, depicted here by Gustave Doré, described Dante's vision of a paradise shining with "many living lights of blinding brightness." (Wikicommons)

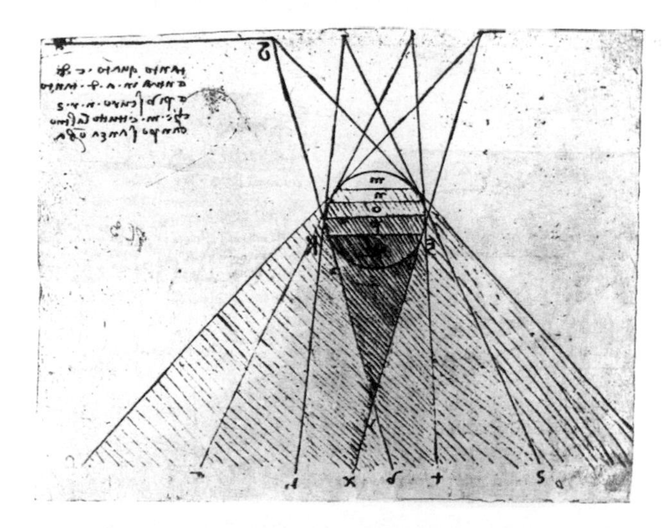

Leonardo da Vinci's sketches enabled artists, who eagerly read his *Treatise on Painting*, to render light's subtle shadows. (Wikicommons)

Isaac Newton's experiments with prisms led to his treatise *Opticks*. Published in 1704, the work laid the foundation for all future study of light. (Getty Images)

"The sun is god," proclaimed J. M. W. Turner, whose *Caernarvon Castle* (c. 1798) heralded a new light in painting. (Public domain)

Discovered in the 1600s, Iceland spar splits a beam of light. The little crystal baffled scientists and inspired new investigation into the wave nature of light. (Julie Kumble)

In 1837, Louis-Jacques-Mandé Daguerre used a variety of chemicals and copper plates to capture the light in his studio. Within two years, Paris was seized by "daguerreotypomania" as thousands bought cameras and made daguerreotypes. (Société Française de Photographie)

Albert Michelson's 1887 experiment disproving the existence of "luminiferous ether" paved the way for Einstein's relativity theory. Michelson, the first American to win the Nobel Prize for Physics, claimed he studied light "because it's so much fun." (AIP Emilio Segrè Visual Archives)

In the spring of 1892, Claude Monet began painting the Gothic cathedral in Rouen in the changing light. Three years later, when twenty cathedral paintings went on display, one critic called them "exasperated and morbid," but another hailed the series as "a revolution without a gunshot." (Public domain)

In his earth-shaking theory of relativity, Albert Einstein used light's constant speed to prove that time is not a constant. But Einstein argued fiercely against the particle-wave duality of light endorsed by Niels Bohr (*left*) and other quantum theorists. (AIP Emilio Segrè Visual Archives)

Each December 21, hundreds gather in Newgrange, Ireland, to see the sun rise on the shortest day of the year. Newgrange's passage burial tomb lights up with the solstice sunrise. (Photo by the author)

By the 1820s, a shop on the Île de la Cité, an island in the river Seine, had cornered the market on all things optical. For Parisians fascinated by light, La Maison Charles-Chevalier was a window of light. The Chevaliers, first Louis-Vincent and later his sons, designed, manufactured, and sold Europe's largest inventory of lenses, spectacles, and other glass trinkets. Browsing the glittering shop that overlooked the arches of the Pont Neuf, astronomers bought the finest telescopes, biologists sampled the latest microscopes, and opera buffs fingered twin-tubes sheathed in *nacre de perle* or *ivoire blanc.* Painters could buy a camera obscura, or its latest offshoot, the camera lucida—a prism and mirror that projected a vertical image onto a horizontal page. La Maison Charles-Chevalier also featured monocles and pince-nez, lenses convex and concave, lorgnettes, mirrors, magnifiers, and more. Among the most popular items was the *lanterne magique.*

Invented in the mid-1600s—some say by wave theorist Christiaan Huygens—the magic lantern was a primitive projector whose candles, lenses, and painted slides beamed images across dark rooms. Flickering ghosts and devils made audiences shudder before this "lantern of fear." Yet the magic lantern taught Parisians another truth about light—that crowds would pay to see it perform.

In the summer of 1798, Parisians began flocking to a light show far more eerie than any "lantern of fear." Filing into a candlelit room, the audience sat in quiet anticipation. A man stepped from behind a curtain. He spoke softly about the supernatural and the need to be skeptical. The man departed; the candles went out. Then without warning, the far wall exploded with bolts of lightning. Skeletons, Medusa heads, and witches floated in the blackness. Specters loomed, growing larger as they danced on the wall. Smoke soon filled the room, filtering light through fumes. The fright show, called *Fantasmagorie*, continued for ninety minutes. Filing out shaken, shuddering, thrilled, the audience spread the word. A few perhaps knew it was all done with a projector, but how had the ghosts danced and rolled their eyes? Was it all illusion? Police did not think so when they briefly closed the show for fear that it might resurrect the guillotined Louis XVI. On into the new century, as the Romantics spread fascination with gothic horror, theaters of *Fantasmagorie* opened in London and New York. By then, Paris was gearing up for its next light show.

Among the regular customers at La Maison Charles-Chevalier was a tall, curly-haired man with large, sad eyes and a cropped mustache. Confident and curious, he most often looked at camera obscuras yet seemed fascinated by all the optical toys. Perhaps as a teenager he had seen the *Fantasmagorie*, and here in the Chevaliers' shop was a home model, wheeled, wooden, with lenses, mirrors, and a tin stovepipe channeling lamp smoke. The price—250 francs—was well within the reach of the dapper customer, but he was not interested. Charles Chevalier knew the gentleman well. A landscape artist by training, the man was designing sets for the Paris Opéra, creating towering curtains painted with arched porticoes and ruins. Though at the height of his profession, the man always inquired about the newest lenses, the latest illusions, the endless possibilities of light. Clerks addressed the man as M. Daguerre.

Before he gave his name to the silvery portraits that pioneered photography, Louis-Jacque-Mandé Daguerre was known throughout Paris as a master of light. Raised in a small town north of Paris, Daguerre began sketching as a child and at the age of thirteen did a portrait of his parents that convinced them of his talent. But art was a dicey business, so Daguerre was apprenticed to an architect. He protested and soon left for Paris. A passable artist, he managed to hang a few paintings in the Paris Salon, yet when hired to paint stage sets, he found a calling. By 1820, having noticed how Parisians loved light, Daguerre had begun manipulating lamps, slides, and curtains to simulate sunsets and moonlit nights. At the Paris Opéra, he painted Mount Etna, backlit the scene, then flared and dimmed overhead lights until the volcano seemed to erupt. The effect, a critic said, was "the most astonishing thing that could be produced." Later M. Daguerre made the sun rise above one set and created a "Palace of Light" whose jewels sparkled beneath fading arches and columns. It was another "marvel by M. Daguerre," who was ready to take stage lighting to a higher level.

On a sunny July morning in 1822, crowds gathered outside the newest auditorium in Paris's theater district. The four-story building featured slim, arched windows beneath huge letters reading:

DIORAMA

At eleven A.M., ushers led crowds through a lobby and into a lamp-lit chamber. All waited to see what new wonder M. Daguerre had devised. Once eyes had adjusted to the dimness, the throng moved into a spacious rotunda with a window on one side. Through the window stood the interior of Canterbury Cathedral, yet the huge stones did not seem in any way artificial. Workmen were doing repairs and many in the audience swore they saw the men move. One woman asked to enter the cathedral. Then the cathedral's stained glass flared and faded. Stones shifted in shadow. As the audience watched wide-eyed, the cathedral receded, growing smaller, fainter, disappearing. Moments later, through the same window, the faint outline of mountains appeared. Brightening, seeming to sketch themselves, the peaks revealed an Alpine valley with chalets, running fountains, and a lake that glittered in sunlight. "The most striking effect is the change of light," the *London Times* wrote. "From a calm, soft, delicious, serene day in summer, the horizon gradually changes, becoming more and more overcast, until a darkness, not the effect of night, but evidently of approaching storm—a murky, tempestuous blackness—discolours every object, making us listen almost for the thunder."

The show lasted just half an hour, but Daguerre's *tableau magique* was soon the talk of Paris. For an admission of two francs, one could watch light sculpt idyllic scenes. Subsequent dioramas presented a Swiss valley during a landslide, the cathedrals at Chartres and Rheims, Venice's Grand Canal, and Napoleon's tomb on Saint Helena. Audiences, fooled again and again, threw paper wads at the painted curtain just to be certain it was all illusion. The press swooned:

- "A veritable triumph."
- "An epoch in the history of painting."
- "Entitles M. Daguerre to be ranked as one of the most distinguished painters that ever lived."

Dioramas opened in London and Berlin. La Maison Charles-Chevalier soon sold Polyorama Panoptiques, backlit boxes with cutaway scenes allowing children to mimic Daguerre's special effects.

Like any magician, Daguerre refused to reveal his secrets. How had he made the moon rise, rivers flow, cathedrals shimmer? Each diorama was painted on a silk curtain using a camera obscura to create the ultrarealistic trompe l'oeil ("trick the eye") that Daguerre had learned as a stage designer. The best-kept secret, however, was the light. Daguerre rendered fixed objects in opaque oils but did windows, streams, candles, and other luminous surfaces in thin, translucent colors or left them unpainted to let light through. His crew, working below the stage, shifted lamps and colored screens, sometimes adding radiance from skylights in the roof. The effects—lighting candles for a midnight mass, shifting the shadows of kneeling figures, then extinguishing candles to darken the church—delighted adults and enthralled children. At one show, a goat appeared in an Alpine village. In the audience were King Louis-Philippe and his young son.

"Papa, is the goat real?" the dauphin asked.

"I don't know, my boy," the king replied. "You must ask Monsieur Daguerre."

More dioramas opened, in Liverpool, Berlin, and Stockholm. Throughout Europe, Daguerre was lauded as a maker of miracles, yet he wanted more from light. He had often spoken with Charles Chevalier in his shop, pointing to scenes in a camera obscura. "Won't one ever succeed in fixing these perfect images?" Daguerre asked. Sometime in 1824, in his home near his diorama theater, Daguerre began experimenting with chemicals and copper plates.

For nearly a century it had been known that silver nitrate, also called "lunar silver," turned gray or black when exposed to light. Silver chloride made an even darker image. The first success in capturing light's patterns on paper dated to the 1790s, when Thomas Wedgwood, heir to the Wedgwood porcelain family, coated paper with silver nitrate, set a leaf on top, put the plate in the sun, then applied oil of lavender to the exposed plate. The result was an exquisite silhouette. None of Wedgwood's "photograms" lasted long, however. Without a fixative, even the dimmest light blackened silver nitrate. Other British scientists tried various chemicals that slowed the darkening—briefly. Then in 1816, a French gentleman on his Burgundy estate took up the quest.

Joseph Nicéphore Niépce preferred silver chloride to silver nitrate. His images were the sharpest yet, but still could not be fixed. After years of frustration, Niépce turned to a curiously named material, bitumen of Judaea. This tarry substance adhered to glass when exposed to light. One morning in May 1826, Niépce filled a wineglass with pulverized bitumen, added oil of lavender, heated the steamy black paste, and layered it on a silver plate. After warming the plate with an iron, he set it in a camera obscura and uncovered the lens. Eight hours later, he closed the lens. Washing the plate to remove the loose bitumen, Niépce held the first photograph. Though blurred by advancing shadows, the photo showed smudged roofs and walls as seen from Niépce's back window. Niépce called his process "heliography," and the term soon reached Charles Chevalier. Daguerre, learning from Chevalier of the breakthrough, wrote Niépce to ask if such a process might be made practical. Niépce sent Daguerre a pewter plate etched by heliography, and in 1829 the two formed a partnership. Niépce died four years later leaving his discovery to his son, who did not share his father's interest. Daguerre pressed on.

Religion had worshipped it, physics had calculated it, art had copied it, and poetry had eulogized it, but it took chemistry to finally seize light.

Daguerre never used bitumen of Judaea. He wanted a perfect image, not some tarry blur. Poring over chemistry books to learn of light-sensitive substances, he tried "Bologna stone," the same glowing mineral that had made Galileo wonder about light. He used iodine vapors to coat silver plates, making silver iodide, which is darker than silver chloride. Carbonic acid reversed negative images, and chlorate of potassium, heated by a lamp, fine-tuned the contrast. One day Daguerre burst into Chevalier's shop. "I have seized the fleeting light!" he shouted. "I have forced the sun to paint pictures for me." Still, the fuzzy shapes would not stop darkening. Daguerre's wife worried that he was "possessed" by his pursuit. "He is always at the thought," Madame Daguerre told a noted chemist. "He cannot sleep at night for it." Was such a dream even possible, she asked? The chemist found it unlikely.

One day in 1835, Daguerre finished his experiments and set a barely exposed plate in his cabinet of chemicals. The next morning, he took out the plate, thinking he would use it again. He stopped, stared. The plate bore a finely etched image, faint but easily the best he had made. Having no idea what had developed the image, Daguerre began setting exposures in the cabinet and removing chemicals one by one. The telltale substance proved to be mercury, its vapors rising from a broken thermometer. More experimentation with fixatives followed, until in 1837 Daguerre made his first permanent image—a smoky still life showing two clay heads on a table. Another eighteen months of work, during which Daguerre tried to get financial backing from foreign governments for his experiments, led to the date that is fixed in the memory of every student of photography.

Before 1839, all the light illuminating all of human history—the rise and fall of empires, the coronations of kings and queens, the lovely faces of children, the wise faces of the elderly—was lost to time. We have only paintings, "which alone," as Leonardo observed, "portray faithfully all the visible works of nature." The pivot point in history's link to light came in Paris at the Institut de France. August 19, 1839. A Monday. A gorgeous summer afternoon in the City of Light. There are no photos of the event, indoor light being far too weak to capture. But picture the dapper, curly-haired Daguerre seated in the ornate hall before a joint gathering of France's Académie des Sciences and Académie des Beaux-Arts. The crowd spills onto the quai overlooking the yellow Seine and the boxy black mansards of the Louvre. This is the moment Paris has awaited since Daguerre's serendipitous accident with mercury vapor was first announced in an arts journal: "It is said that M. Daguerre has discovered a means of receiving on a plate of his own preparation the images produced by the camera obscura. Physical science has probably never presented a marvel comparable to this." But four years have passed without proof, and skeptics have come creeping. "The wish to capture evanescent reflections is not only impossible," a German newspaper warned, "but the mere desire alone, the will to do so, is blasphemy. God created man in His own image, and no man-made machine may fix the image of God."

In January 1839, hoping to sell the French government the rights to his process, Daguerre had described it in a public broadsheet. Though he pinned his own name to the invention, he credited Joseph Niépce with the breakthrough, claiming only to have sharpened images and cut exposure time to three minutes. Portraits would not be possible at such a speed, Daguerre warned, but he foresaw people making daguerreotypes of their homes, scientists imaging the moon, travelers preserving the landscapes of their dreams. "The daguerreotype," he concluded, "is not merely an instrument which serves to draw Nature; on the contrary it is a chemical and physical process which gives her the power to reproduce herself."

But how was such a miracle performed? Daguerre refused to say. To drum up support, he spent the spring of 1839 lugging his hundred-pound camera to the Louvre, Notre-Dame, and other landmarks. Presenting postcard images to government officials, he asked two hundred thousand francs for all rights. No government had ever bought an invention outright, however, and politicians balked. By June, the public's initial enthusiasm had waned, leaving Paris again skeptical. As the inventor of the diorama, Daguerre was, after all, an illusionist. Then in July, François Arago, champion of Fresnel's proof of wave theory, threw his full weight behind Daguerre. The distinguished astronomer, now a member of France's Chamber of Deputies, convinced his colleagues. By nearly unanimous vote, the chamber approved the government's purchase of "the eye and pencil of this new painter." Daguerre, now in his fifties, had lowered his monetary demand to a modest pension. France awarded Daguerre six thousand francs annually; Joseph Niépce's son got four thousand.

A month later, on August 19, the veil was finally lifted. Daguerre, nervous and shaking, pleaded a sore throat, and sat in silence while François Arago called the overflow audience at the Institut de France to order. Standing before a hall packed with luminaries, he explained this latest triumph of science and of France. Before he finished, crowds outside were shouting the secrets along the quai.

"Silver iodide!"

"Quicksilver!"

"Hypo-sulphite of soda!"

Within the hour, La Maison Charles-Chevalier sold out of camera obscuras. Seizing the light, however, was not as simple as clicking a shutter. Arago's explanation left many confused, forcing Daguerre to demonstrate later for reporters in his studio. Taking out a copper plate coated with silver, he buffed it with powdered pumice, then rinsed it in a solution of nitric acid. After darkening the room, he placed the plate facedown on a box containing heated iodine. The vapors, filtered through muslin, turned the plate yellow. Daguerre then fit the plate into his camera, a metal box two feet long and one foot square with a lens ground by Charles Chevalier. Hefting the camera to his window, Daguerre looked at his watch. It was a gray afternoon. The exposure time, he announced, would be shorter on a sunny day. Or in Spain or Italy. He uncovered the lens. Outside, carriages, horses, and people passed, but their fleeting images would be only faint blurs on his plate. Daguerre checked his watch again. Reporters fidgeted. After several minutes, he covered the lens, removed the plate, and leaned it above a tin filled with mercury. He set a flame beneath. When the mercury reached 60 degrees Celsius, Daguerre let the steaming quicksilver cool, then took out the plate and bathed it in a solution of sodium thiosulfate (the "hypo-sulfate of soda" called out on the quai). The entire process took a half hour. Reporters huddled around as Daguerre held before them a crystalline image, five inches square, of what the sun had painted outside his window. Through glasses, monocles, and pince-nez, reporters examined each intricate detail. Eyebrows rose. The wonder had begun.

Paris quickly succumbed to "daguerreotypomania." Throughout the city, men hefted bulky boxes and balanced them on tripods. Iodine and mercury sold out in pharmacies. "We all felt an extraordinary emotion and unknown sensations which made us madly gay," one man recalled. "Everyone wanted to copy the view offered by his window . . . [and] the poorest pictures caused him unutterable joy." By mid-September, daguerreotypes were made in London and New York. Daguerre's manual on the process sold through eight printings in several languages. Men with cameras and chemicals set out for Italy, Egypt, and the Holy Land. By the time they returned, two French physicists had made daguerreotypes of the moon and through a microscope. A British chemist, adding chlorine vapor to the stew, reduced exposure time to a few seconds. Soon the well-heeled and well-known were seated before cameras, their necks held by braces. Within a few minutes, each left the daguerreotype studio holding a lustrous portrait.

The earliest daguerreotypes are not so much photographs as small wonders. Etched in silver iodide, they show the Parthenon, the Leaning

Tower of Pisa, and the Roman Forum, empty and abandoned as it appeared in 1840. Then come the famous first portraits. France's King Louis-Philippe. England's Duke of Wellington. Lincoln. Whitman. Poe. But many early daguerreotypes show ordinary people—stern women in frilly bonnets, stalwart men in uniform, children frozen in time. And one shows Daguerre himself, plump and rigid, posing for the process that would evolve into photogravure, lithographs, celluloid film, Kodak cameras . . .

In 1840, with his pension and fame fixed, Daguerre retired to a small town outside Paris, emerging only to make a small diorama in a local church. (The diorama of a cathedral interior, recently restored, is on view in the little white church of Saint Gervais-Saint et Saint-Protais in Bry-sur-Marne.) Daguerre had seized the light, and never again would it seem so fleeting. Once his process was simplified, photography put light into human hands. The initial frenzy of "daguerreotypomania" has barely slackened. Some 3.5 trillion photos have been taken since 1839, and the spread of smartphones has upped the numbers exponentially. Today, nearly four hundred billion photos are taken each year, more snapped every few minutes than were taken during the entire nineteenth century.

The American artist and inventor Samuel Morse, visiting Daguerre in Paris, called his work "Rembrandt perfected." Other artists were not so sure. "From today," one observed, "painting is dead." Yet France was just starting to shine.

By 1848, the City of Light had become a city of gloom. Even as daguerreotypists focused on the city's treasures, nearby tenements made Paris "a great manufactory of putrefaction in which poverty, plague . . . and disease labor in concert and where sunlight barely ever enters." The vast majority of the one million Parisians lived on half a franc per day, not enough to buy a baguette. Early in that landmark year, corruption and scandal led to another uprising. Again the barricades went up, again the blood flowed, and the regime of Louis-Philippe fell. Supplanting the monarchy came another Napoleon, nephew of the deposed Bonaparte. Though elected by popular vote, Louis-Napoleon soon dissolved the Chamber of Deputies and, in 1852, declared himself Emperor Napoleon III. Because dank, narrow streets were breeding grounds for revolt, the emperor began making plans for gutting Paris.

Throughout the 1850s Paris was rebuilt and reborn. Broad boulevards tore through former slums. New squares and expanded parks—the Bois de

Boulogne, the Bois de Vincennes—let in light. Sewers that had emptied into the Seine were rerouted. A third of the population was displaced, whole neighborhoods were razed. Nearly twenty thousand buildings fell and thirty-four thousand new structures rose. Critics denounced the upheaval and its mastermind, Baron Georges-Eugène Haussmann, but when the project was finished, Paris finally earned its nickname. Now, its *grands boulevards* shone beneath open skies. Light was everywhere in the city, glinting from chandeliers, beaming from cafés, blazing from fifteen thousand gas lamps lining the streets. In 1855, when Paris held the first of many expositions, a Fresnel lighthouse lens shone above the Champs-Elysées. Light—free, abundant, joyous light—began to change how Parisians saw the city, the sky, and each other. A decade later, it was time for artists to catch this light.

Edgar Degas caught the stage footlights on a dancer's diaphanous skirt. Berthe Morisot caught light falling on lace above a baby's crib. Auguste Renoir caught summer light dappling shaded picnics. Edward Sisley caught winter light on snow. Georges Seurat caught every point of light (and then some) on a Sunday afternoon on the Island of La Grand Jatte. And Claude Monet caught haystacks in the soft grays of morning, the orange blaze of noon, the violet shadows of evening . . . But whether painting dancers or haystacks, seascapes or snowscapes, the subject was light. "The motif for me is nothing but an insignificant matter," Monet said. "What I want to reproduce is what there is between the motif and myself."

"Drunk on sunshine," as Renoir described them, the Impressionists smashed art's tired rules. Leonardo had told painters to work in shadow, yet Impressionists, when not painting in pure sunlight, floodlit their studios. Caravaggio had made chiaroscuro a staple of Western art, and his showy luster a fixture of still life. The Impressionists disdained shadows and luster, preferring a light that came from everywhere and nowhere. Artists, Leon Battista Alberti had advised, should use plain white sparingly. Yet the Impressionists were drawn to snowy fields, frothy oceans, white flowers in a white vase, women in white chenille on white blankets. And while artists for centuries had reveled in black, Monet dismissed the tradition in a single phrase: "Black is not a color."

The allure of Impressionist light is familiar to anyone who visits a major museum. Room after room hangs heavy with medieval gold leaf or Renaissance ochre. Next come galleries filled with the somber tones that the Impressionist Edouard Manet mocked as "brown gravy." The light of paintings prior to Impressionism resembled the *lux* of Roman and biblical days—focused and powerful—streaming through windows, glinting off objects, bursting from behind clouds. But Daguerre had shown *lux* to be just an artist's conception. The light frozen by daguerreotypes was not *lux* but *lumen*, an ambience that might sparkle off a lake but spread evenly over land and sky. To paint such light, artists would have to forget Caravaggio's glare and Rembrandt's glow. And once any fool could take a photograph, art demanded more than pictorial perfection. Impressionists set out to portray light not as a camera could but as the human eye saw it—fuzzy and fleeting. "Their aim," wrote the art historian Anthea Callen, "was to perceive and record direct optical sense data, or 'visual sensations' . . . This explains references among the Impressionists to the desire for the infant's untutored eye." The revolution was sudden, shocking, and, unlike previous French revolutions, involved not barricades but paint and canvas.

Regardless of their trademark styles, the Impressionists shared certain techniques. Each tired masterpiece in the Louvre gave off its own light, reflected in varnish and oils, yet Impressionist works did not glow. To mute the sheen of ordinary oils, they squeezed paint onto blotting paper, letting it bleed until the colors were matted. Turpentine could also dampen glare, and Monet sometimes mixed a chalky paste into his paint. Classical artists had used bitumen of Judaea, the same tarry substance that made the first photograph, to darken shadows. But how could you darken light? Manet had the answer, using thick brushstrokes to enhance the opacity of any color. Leonardo had achieved his sfumato by coating paintings in grainy varnish, but the Impressionists refused to varnish finished works. Artists since the Renaissance had primed bare canvases with dark layers, but the Impressionists, following Turner, used a base of pure white.

The light they caught was the light Paris saw, yet the Impressionists endured a decade of mockery. Rejected by the Paris Salon in 1864, they hung their works across the street in the Salon des Refusés. "People entered it as they would the horror chamber at Madame Tussaud's in London," one artist recalled. "They laughed as soon as they had passed the door." Queen Victoria had thought Turner mad, yet Parisians just considered the Impressionists obscene, and their works "the ultimate in ugliness." Still, these outcasts were drawn to the fugitive light. Studying the latest color theory, meeting in cafés, painting together outside, *en plein air*, sipping absinthe, loving women, romanticizing everything they touched, the Impressionists gradually wore down resistance and opened eyes. This, after all, was what Paris *looked* like. Daguerre had frozen its light on copper plates; the Impressionists did the same on canvas. And finally, grudgingly, "the new painting" took hold. A dozen years after the Salon des Refusés, the author of a widely read pamphlet wrote of these renegades, "Proceeding from intuition to intuition, they have little by little succeeded in breaking down sunlight into its rays, its elements . . . The most learned physicist could find nothing to criticize in their analyses of light."

By the mid-1890s, Impressionism had converted all but the stuffiest critics. In May 1895, crowds lined up outside the Durand-Ruel Gallery, a few blocks north of the Louvre, to see the latest show by the now-famous Claude Monet. Of the show's fifty canvases, twenty were of the same subject—a cathedral in Rouen. Monet had spent two successive springs living alone in Rouen, painting in a small apartment and a nearby shop. Each venue opened onto the square shadowed by the great Gothic cathedral, whose construction had begun shortly after its bishop saw Abbot Suger's light in Saint-Denis. At first Monet adored the old cathedral. Lavishing his matted colors on its massive towers and rose window, he worked on several canvases at once, painting the light as it changed with the hour. But he soon came to loathe every stone. Frustrated by damp weather, enraged by the vicissitudes of sunlight, he obsessed over the project. "Imagine that I get up before 6 A.M. and at work from 7 A.M. until 6:30 in the evening," he wrote his fiancée. "Always standing—nine canvases. It is killing, and for this matter I abandon everything, you, my garden." He destroyed some canvases, started others—*Morning Effect, The Facade in Sunlight, Setting Sun (Symphony in Gray and Pink)*. Falling exhausted into bed, he suffered nightmares of the cathedral in a hideous green, or else crashing down on him.

Five weeks into his first spring in Rouen, Monet abandoned the project, packed his canvases, and returned to Giverny to get married. All that summer he painted his garden and other subjects, letting the radiance of Rouen go untamed. But the following February he was back before the cathedral. The weather was worse, his progress "heartbreaking." "My goodness!" he wrote to his new wife, "they are not very far-sighted, those who see a master in me! Beautiful intentions, yes, but that's it! Happy the younger ones, those who believe that it is easy. I was like that once, but it is over; yet tomorrow morning at 7:00 I'll be there." He continued painting on into March—drizzle, sunshine, more drizzle. His left thumb, stuck all day in his palette, grew numb and swollen; his back ached. He was in his early fifties, and a prisoner of his vision, of light. But then spring released him. Suddenly Monet saw "no longer the oblique light of the February days; it is every day more white, more vertical, and as early as tomorrow I am going to work on two or three extra canvases." Finally, in mid-April 1893, the series was done. Monet spent another two years adding final touches, and the show opened.

Crowds wandered through the Durand-Ruel Gallery, intrigued, surprised, not knowing what to say. Here was the same cathedral, its soaring arches and porticoes on canvas after canvas. Only the light was different—soft blue when caught in early morning, golden at noon, gray in the afternoon, white in the fog. Critical reaction was mixed. One called the cathedrals "exasperated and morbid," but another hailed the series as "a revolution without a gunshot." Later art historians have seen the cathedrals as "the last thing that nineteenth-century French painting had to say." What Monet's cathedrals said, however, continued into the next century. By then, younger artists were fracturing light's pyramids, exaggerating colors, refusing to admit that there had ever been any rules.

France, while careening from monarchy to revolution to empire and back, had given light a new role—as a time machine. Sunlight had always been a reliable clock, but now the clock could be stopped. Now a moment could be fixed on copper plate or on canvas. Daguerre, catching the view from his window and showing it to reporters, held a piece of afternoon. Monet, painting Gothic facades subtitled *Morning 9–10 A.M.* and *Late Morning 11:45 A.M.–12 Noon*, made time, as much as light, his motif. This newly tamed light was a gift to those who could hold it, a gift we still open whenever a digital camera or smartphone is passed around so all can admire the scene shot seconds ago. Not much has changed in those seconds. Anyone present could turn and see the same scenery, the same faces, larger and alive. Yet all are captivated by the image. Here is how the light looked, here in your hand. "Precisely by slicing out this moment and freezing it," Susan Sontag observed, "all photographs testify to time's relentless melt."

Paris, December 28, 1895, 14 Boulevard des Capucines: A block from the Paris Opéra, men in top hats and women in flowing dresses swarm along the open street. Outside the Grand Café, a man beckons strangers to see the latest wonder of light, the *Cinématographe.* Ever interested in light, a few dozen Parisians pay a franc each, descend a circular staircase, and take their seats. A white sheet hangs on the wall. Behind the audience sits a wooden box with a protruding lens. Kerosene lamps go out. A single image appears on the sheet, a frozen image of workers standing at the edge of a factory. A photograph? Is this the latest spectacle? Then the man from outside turns a crank on the box. The light flickers. Figures on the sheet begin moving! Walking! Slowly at first, then at normal speed, women in long dresses, men in suspenders and straw hats hustle out of the factory, turning right or left and disappearing. A dog wanders by. A man rides a bike. A child scurries past. The moving images last less than a minute, but after a few dark moments, another *actualité* begins. This time a train approaches, the enormous engine heading straight for the audience. Several leap to their feet; a few bolt for the door.

Motion pictures—moving light—have arrived. The inventors are two brothers from Lyon, Auguste and Louis Lumière. The Lumière Cinématographe on the Boulevard des Capucines is soon raking in seven thousand francs a week. Paris finds it ironic that *lumière* means "light."

CHAPTER 12

"Little Globe of Sunshine": Electricity Conquers the Night

This crystal tube the electric ray
Shows optically clean,
No dust or haze within, but stay!
All has not yet been seen.

—JAMES CLERK MAXWELL

Before it arrived with the flick of a switch, making light was less a puzzle than a nuisance. First the nearest animal—deer or cow, yak or ox—had to be slaughtered and stripped of its fat. Soon kitchens and huts reeked with the stench of unctuous, oily suet boiled in vats of quicklime. As the mixture thickened, sulfuric acid decomposed the lime, leaving a greasy slurry. Next, strings dangling from wooden rods were dipped and dipped again until a finger of white wax clung to each wick. Now the energy stored in a dead animal, cooled and cut into candles, was ready to burn. Strike a chunk of flint with a stone, strike until a spark caught the wick and the blessed light flared. An alternate home recipe for light used animal fat or flammable oil poured into lamps—lamps of stone, then of clay, later of glass and pewter, wood and porcelain. And so the centuries passed, millennia of nights lit solely by candles and lamps.

Only in the 1700s did better light emerge. Oil lanterns with double wicks. Then, late in the century, the spermaceti from certain whales was found to burn twice as bright as other fat, sending men to sea in ships. Next came Argand lamps, whose hollow wicks fed air into a flame as bright as six candles. But Argand lamps were found only in Europe and its colonies. On into the 1800s, gas lamps lit city streets and mansions, yet in huts and houses and villages around the globe, dusk kindled the light of the ancient world. Kerosene, when refined from oil wells in the 1860s, offered some relief from the grease, the smell, the slaughter. But light still demanded flame and flame brought smoke and sometimes fire. Simply tipping over a candle or lamp could burn down a home, or half a city. London. Moscow. Amsterdam. Tokyo. Boston. Chicago . . . Their "great fires" were the price light exacted in return for battling the "horrid darkness."

Throughout the millennia of candles and lamps, the source of light that would one day flood the world lay hidden in plain air. Only the curious considered it. The Greek philosopher Thales saw sparks fly from chunks of amber. Chinese Mohists also observed static electricity, as did the Vaisheshika school in India. Yet almost two thousand years passed with no way to capture these sparks. Not even Newton could imagine electricity as anything but "a certain most subtle Spirit, which pervades and lies hid in all gross bodies . . . that cannot be explain'd in few words." Sparks could make men jump and women's hair stand on end. Trapped by the electrodes of a Leyden jar—an eighteenth-century invention for creating static charges—electricity could release a bolt that sent a man flying across a room. Surely a force so vagrant, so dangerous, could never become a steady source of light.

Light and electricity seemed destined to remain as tangential as wind and water. When the British chemist Humphrey Davy, in 1802, sent a small current through a strand of platinum, a searing glow fried the filament in less than a second. Seven years later, Davy created a different electric light—the arc lamp, its spark leaping a gap between carbon rods. But arc lamps were too blinding for home use, and when finally made practical for city streets, still scorched the eyes. "A new sort of urban star now shines out nightly," Robert Louis Stevenson lamented in 1878, "horrible, unearthly, obnoxious to the human eye; a lamp for a nightmare! Such a light as this should shine only on murders and public crime, or along the corridors of lunatic asylums." But even as Stevenson made his "plea for gas lamps," inventors were racing to patent small "incandescent lamps." Skeptics abounded. The *New York Times* quoted "sober scientific electricians [who said] that while large spaces can be conveniently and economically lighted by electricity, bedrooms, parlors, and small spaces requiring the light of only the power of a few candles cannot be so illuminated." Household light had to be steady and dependable, yet electricity was as evasive as an eel, as deadly as lightning. The two were finally linked only by the inventiveness of two kind and congenial men.

With Michael Faraday and James Clerk Maxwell, the narrow paths paved by light's earlier students began to broaden. The self-absorption of the pre-Socratics, the monastic focus of Robert Grosseteste and Roger Bacon, the bitter rivalry of Newton and Hooke, the contests between Fresnel and the French Académie—each phase had run its course. There would still be the usual academic jousting, and the fierce competition of those trying to make fortunes from light, but on into the twentieth century, physicists examining "the thing you call light" would share their discoveries. Challenging rivals' findings but not their integrity, complementing each other's work, meeting for long walks and longer conferences, they would add, piece by piece, to the puzzle. Such cooperation might signal the maturation of science, or merely its divergence from philosophy, yet the most likely cause was complexity. The more learned about light, the stranger it would seem. The math would get harder, the quirks quirkier, suggesting that if light was ever to be understood, the task would take all the insight of all the best minds, working together. From here on in, those who studied light would be, with a few exceptions, gentlemen scientists. The turn away from ego began with Michael Faraday.

Isaac Newton had lectured "to the walls," but Michael Faraday's talks in London drew packed audiences. None of the Victorians who came to hear his Friday Evening Discourses were students, because Faraday was no professor. A self-taught grade-school dropout, Faraday used his boyish enthusiasm to impress scientists, and his penetrating intelligence to lift the veil that shrouded the body electric. Standing just five feet four, with a wild shock of hair and an impish gleam in his eye, Faraday was the science teacher every student dreams of—patient, clever, and eager to make ordinary objects ignite. He began each discourse by lighting a small flame on the table before him, a tradition that continues at London's Royal Institution, where the weekly Friday Discourses that Faraday initiated still take place. Faraday began them in 1826, and later added annual Christmas lectures for children, speaking to them as fellow "philosophers." By 1850, Michael Faraday was as much a London celebrity as Charles Dickens. The novelist George Eliot found Faraday's talks "as fashionable an amusement as the Opera." Others were drawn by his joy, which "seemed to carry him to the point of ecstasy when he expatiated on the beauties of Nature."

The son of a blacksmith in London's South End, Faraday was an apprentice bookbinder when he was hired by Humphrey Davy. The inventor of the arc lamp simply needed a secretary, but Davy found that Faraday had attended his chemistry lectures, taken meticulous notes, and bound them for further study. Faraday was given an attic room at London's Royal Institution plus coal, candles, one daily meal, and twenty-five shillings a week. All he knew about electricity was what he had read in the *Encyclopedia Britannica*, but after helping Davy perfect the next improvement in lighting—a sealed miner's lamp that would not ignite methane fumes—Faraday was working in the lab one day in 1820 when amazing news came from Denmark. While lecturing on electricity, the physicist Hans Christian Øersted happened to hold a coil of wire, attached to a battery, near a compass. The needle moved. The needle moved! Could electricity be the same force as magnetism? Øersted soon came to believe that "all phenomena are produced by the same power"; by "phenomena" he meant heat, electricity, magnetism, and light.

Hearing the news, Davy and his young apprentice quickly wrapped a coil around an iron bar and hooked wires to a battery. Again, the needle moved. But Davy soon grew jealous of Faraday—the great chemist would later say that his own biggest discovery was not sodium or calcium, but Michael Faraday—and the two men dodged each other until Davy's death in 1829. From then on, Faraday devoted his life to electricity and magnetism. He knew little math, and often wished that physics could be translated "out of hieroglyphics." But his "want of mathematical knowledge" made him insist on observing rather than calculating nature. Throughout the 1830s, Faraday tinkered with primitive coils, batteries, and voltage meters to lay the groundwork for the study of electricity. He made the first motor, a simple needle dangling above an electromagnet standing in a vat of mercury. Connecting a battery, Faraday and his wife's younger brother watched as the needle spun and spun. The boy would recall his uncle dancing around the room. Later, thrusting a magnet in and out of a wire coil, Faraday saw his voltage meter go wild. He had created the first electric generator. He then began pondering a radical idea about how the forces of nature travel through air or ether.

Devoted first to chemistry and then to electricity, Faraday had paid little attention to light. But hearing of Fresnel's work, he took notice. Though baffled by Fresnel's math, Faraday was intrigued by waves and polarized light. He wondered if the latter might, like a magnet, be drawn to an electric current. In 1822 he bounced light off a mirror, then sent the polarized beam through various liquids hooked to a battery. He hoped to see the beam shift, but it traveled as straight as any shaft aimed by Newton. Faraday remained suspicious. Light, electricity, magnetism—all were swift, ethereal, undulatory. Though long considered to be separate "imponderable fluids," they were, Faraday was certain, closely related.

Fresnel's waves also suggested a new cause of motion. Ever since Newton had laid down his laws, force was thought to travel one of two ways. Objects were either moved by direct contact—tennis balls or particles—or else by a force acting at a distance through the ether. Waves offered a third way—a field: concentric "lines of force" revealed by the pattern iron filings make near a magnet. In March 1832, Faraday scribbled a note, dated it, and stuck it in a safe. Twice accused of stealing others' ideas, including his pioneering electric motor, he wanted to preempt further suspicion. The note described his evolving belief that electricity was not linked only to magnetism but "most probably, to light." Leaving the idea in his safe, Faraday spent several more years studying magnets and electricity. When he returned to light, his discovery would trigger the end of eons lit only by candles and lamps.

Late in the summer of 1845, Faraday set up his most complex experiment yet. An oil lamp. A prism to polarize its light. A chunk of glass so thick that a beam barely passed through it. And beside the glass, an iron horseshoe wrapped with wire, an electromagnet so hulking that it was delivered in a horse-drawn cart. Standing at his lab table, Faraday aligned a prism to catch the beam bent by the thick glass. He still believed polarized light would shift in a magnetic field, yet early attempts—with small magnets, thin glass, even Iceland spar—had failed. Faraday wrote a friend, "Only the very strongest conviction that Light, Mag, and Electricity must be connected" led him to try one more time, with the biggest magnet on earth.

Now he lights his lamp. Now he peers through his lens, adjusting its angle. The light fades and disappears. The lens is blocking the polarized beam. Now he connects the hulking magnet to a battery. He looks through the lens again—and there is the beam. Slightly, ever so slightly, the light has twisted. He repeats the experiment, hooking and unhooking the huge horseshoe. When the current is off, the lens blocks the light at an angle perpendicular to the table, but when the current is on, the light shines through. The angle of polarization has been changed. Changed by a magnet. After further experiments, with a stronger magnet twisting the beam still further, Faraday shocked the Royal Society in 1846, announcing that he had "succeeded in . . . magnetizing and electrifying a ray of light."

Light is magnetic? Not exactly. When passed through a magnetic field, a beam of light does not veer as it does through water or glass. What Faraday saw was how polarized light's spiral football, the verticality of its wave, is torqued by magnetism. Imagining donning sunglasses and turning them until they block the light from your laptop. Imagine getting some horses to drag in a five-hundred-pound electromagnet. Set it up beside your screen, turn it on, and you'll have to shift the angle at which your glasses filter the laptop's light. (You will also need a new hard drive.) Faraday predicted that the effect "will most likely prove exceedingly fertile."

An entry in Faraday's notebooks summed up the man. "ALL THIS IS A DREAM," he wrote. "Still, examine it by a few experiments. Nothing is too wonderful to be true, if it be consistent with the laws of nature." What is now called the Faraday Effect was hailed as the first hint that light might be electromagnetic energy. But light's students, like all physicists, demand mathematical proof. Could some human calculator do for Faraday what Fresnel's integrals had done for wave theory?

In 1857, a dozen years after his magnet twisted light, Faraday received a treatise from a young professor in Scotland. Faraday cringed as he read the title—"On Faraday's Lines of Force." Here was his own work, laid out in equations. "I was at first almost frightened when I saw the mathematical force made to bear on the subject," Faraday wrote back, "and then wondered to see that the subject stood it so well." The author of the treatise, James Clerk Maxwell, was a twenty-six-year-old physics professor at a little-known college in the far north of Scotland. Though he had never attended Faraday's Christmas talks for children, Maxwell embodied their spirit. In Faraday's most famous Christmas lecture, "The Chemical History of a Candle," he had told the children, "We come here to be philosophers, and I hope you will always remember that whenever a result happens, especially if it be new, you should say, 'What is the cause?' 'Why does it occur?'"

Maxwell, raised in comfort on a six-thousand-acre Scottish estate, had delighted his parents by inquiring of the nature around him, "What's the go o' that? . . . what's the particular go of it?" Maxwell's earliest memory was of sunlight. Given a tin pan, the bright-faced boy made it into a mirror, shining sunlight on the ceiling. "It's the sun!" he shouted. "I got it with the tin plate!" Though Maxwell published his first mathematical paper when he was just fourteen, he was equally enthralled with poetry. He would write lyrical verse for the rest of his brief life, sometimes extemporizing while others watched. Once, blending physics with the verse of Scotland's beloved Robert Burns, Maxwell wrote:

Gin a body meet a body
Flyin' through the air
Gin a body hit a body,
Will it fly? And Where?

While studying at Cambridge, Maxwell gravitated toward wave pioneer Thomas Young's tricolor theory of vision—that all colors could be made from just three hues. Ever at play in the fields of physics, Maxwell glued tinted paper onto a child's top, spun it, and saw a rainbow of colors. Adding a sliding dial to change the percentage of each color, Maxwell found he could make a myriad of shades. A few years later, he put Young's theory to more practical use when he made the first color photograph. Maxwell's discovery was surprisingly simple. To make the many colors mixed by just three, he shot photographs of the same object through filters of red, green, and blue. Superimposing the images through magic lantern slides, he amazed audiences during a Friday Evening Discourse at London's Royal Institution. There on the wall was a full color image of a Scottish tartan. Color photography would not become common for another eighty years, but James Clerk Maxwell would get used to being ahead of his time.

Deeply religious, shy with strangers, charming around friends, Maxwell loved literature and philosophy, but he seemed to *see* physics. He could *see* how electricity flowed, how magnets attracted, how light traveled, and how these three "imponderables" behaved as one. He often worked for months on a paper, then left it for months more, working, as he said, in "the department of the mind conducted independently of consciousness." When he finally returned to his subject, he finished with a dash of utter originality. Along with making the first color photo, Maxwell proved that Saturn's rings cannot be solid but must resemble a "flight of brick-bats." He calculated how gases expand and contract, and for fun, the path a sheet of paper takes when floating to the floor. He lamented that he was sometimes "a calculating machine," yet in light's pantheon, only Newton was a better human calculator. Whether walking on his estate, strolling the streets of London, or lecturing in a classroom, Maxwell was subject to sudden insight. At such moments he could "feel the electrical state coming on." Late in 1862, the electrical state came on and, for once, did not let him rest. "I also have a paper afloat," he wrote a friend, "containing an electromagnetic theory of light, which, till I am convinced to the contrary, I hold to be great guns." When the paper was finished, this polite and bashful Scot, heir to the interests of *phusikoi* and philosophers alike, had become the first to *see* the braided wonder of light.

Atoms or eidola. Corporeal or spiritual. Particle or wave. All engaged their eras, but Maxwell ushered in a new era. Picking up where Faraday left off, he began calculating how electricity and magnetism intertwined. By 1863, the "calculating machine" had devised twenty equations, later simplified to four. Maxwell's equations, while as complex as Fresnel's integrals, approached the elegance of $E = mc^2$. Einstein would later pay tribute: "One scientific epoch ended and another began with James Clerk Maxwell." Maxwell's equations proved that electricity and magnetism are tightly bound in the fabric of nature, that they follow Faraday's "lines of force" in a field, and that changes in one field create changes in the other, depending on the conductivity of a material, density of atoms, and a few other factors. Taken together, Maxwell's equations paved the way for the electric age to come. They also opened the box that had trapped light since the first creation stories.

Reading of Faraday's huge magnet twisting a polarized beam, Maxwell suspected that light was electromagnetic. The clues were more than coincidence. Both light and electricity followed the inverse square law, weakening exponentially with distance. Both traveled in waves, not in sweeping longitudinal waves as did sound, but in transverse waves whipping side by side as they advanced. Might the speed of those waves be the key to the connection? Maxwell spent a summer at his Scottish estate pondering the speed of electricity and light. Having left his books in London, he could only ponder. When he returned in the fall, he made more calculations, clocking electricity at 310,740 kilometers per second. This was slightly high yet near the most recent speed of light measured in Paris in 1850. How could such an unfathomable speed be a coincidence? More equations followed, and in December 1864, Maxwell stood before London's Royal Society to present "A Dynamical Theory of the Electromagnetic Field."

Before he reached page two, Maxwell had proposed a sea change in our understanding of electricity, magnetism, and light. They were not separate but were manifestations of a single force. Maxwell went on, spinning out terms—flux, density, displacement, curl—to explain how each force traveled through "an aethereal medium." Finally, noting the speed of electricity, he said, "This velocity is so nearly that of light, that it seems we have strong reason to conclude that light itself (including radiant heat, and other radiations if any) is an electro-magnetic disturbance in the form of waves propagated through the electro-magnetic field according to electro-magnetic laws." The audience sat in silence. Maxwell continued. What followed was Fresnel's math taken to the nth power. Using the Arabs' algebra, Huygens's wave theory, Newton's calculus, the incipient field of vectors, and his own "electrical state of mind," Maxwell laid out fundamental proof. Light was not some mysterious effluvium following its own peculiar laws, but was part of a continuum—the electromagnetic spectrum.

The suggestion was not entirely new. Suspicion that there was more to light than excited the eye dated to 1800, when astronomer William Herschel, setting a thermometer inches beyond the red end of a spectrum, saw its mercury rise. Heat, Herschel posited, consisted of invisible "Calorific Rays"—what is now called infrared light. A few years later, a German chemist saw silver chloride turn purple when set in the darkness beyond the other end of Newton's spectrum, the violet end. Ultraviolet light. Now Maxwell was saying that all light was part of a spectrum—electricity and magnetism interwoven.

Maxwell's light—all light—is not one wave but two. The first wave carries an electrical component of a light beam, the other its magnetic part. A beam's electrical and magnetic parts travel in what Maxwell called a "mutual embrace." One wave snaps vertically as it advances, the other slithers at a 90 degree angle to the first. Together they resemble a roller coaster and its shadow. So what is light? It's the most devilish rope trick ever invented. To play, Nature whips one rope up and down while snaking another along the horizontal plane. The wave crests remain in sync as they advance at 186,000 miles per second. The distance between the peaks (wavelength) determines the color, with red being longer waves, blue and violet shorter. And the higher the peaks (amplitude), the brighter the light. The wonder is not that it took so long to figure this out. The wonder is that it was ever figured out.

Just as there would be no color photography for decades after Maxwell, his "Dynamical Theory" was too complex for his peers. None of the Royal Society members—top scientists in the most dominant country on earth—understood Maxwell's math well enough even to challenge it. William Thompson (Lord Kelvin), later celebrated for formulating the laws of thermodynamics, thought Maxwell had "lapsed into mysticism." The genial Maxwell did not bother to argue. He went on to study other topics—more on colors, on gases. In 1865, he and his wife moved back to his estate in Scotland. There he walked his dog, Toby, sometimes sharing his ideas with the Scottish terrier. Maxwell grew a bushy black beard, and checked and rechecked the speed of light. Several years later he returned to Cambridge to set up the Cavendish Laboratory where the electron, DNA, and dozens of other Nobel Prize–winning discoveries would be made. Though saddled by administrative duties, Maxwell found time to write a textbook on electromagnetism—it ran to a thousand pages. Some speculate that had he lived longer, he would have gotten the jump on Einstein and relativity, but in 1878 he was struck by crippling stomach pains. He had contracted the abdominal cancer that had killed his mother in middle age. He died at forty-eight. The year was 1879. Maxwell's equations had been mastered by perhaps a dozen people on earth, yet one did not need to understand electromagnetism to know that making a safer, cheaper light could make someone very rich.

The quest for a household incandescent light dated to the late 1830s. Inventors from Belgium, England, France, Germany, Russia, and later from America, had labored to find a filament that would burn without burning out. Using primitive batteries, they ran small currents through pencil-thin shafts of carbon, platinum, iridium, paper, cardboard, even asbestos. But bulbs cracked. Platinum fused with wires. Filaments flared out or blackened glass with soot. The British inventor Joseph Swan, a Victorian gentleman with a Santa Claus–white beard, patented a bulb in 1860. But frustrated by one brief flash after another, Swan turned to other inventions. By 1878, he was again testing filaments and hoping to have a bulb ready for public display by Christmas. Then, that spring, in a cluttered laboratory on New York's teeming Lower East Side, a big, slovenly man fine-tuned a carbon coil in a flask filled with nitrogen. Hooking wires to a battery, William Sawyer connected the two. "I tried the full power on the lamp yesterday," Sawyer wrote to his financial backers on March 7, 1878, "and you would have supposed a small sun was shining in the vicinity."

Sawyer, having already earned one patent for an incandescent bulb, filed for another. His breakthrough put him ahead of England's Joseph Swan and America's odds-on-favorite, Thomas Edison. Two years earlier, shortly before astounding the world with his phonograph, Edison had experimented with incandescence but had given up, finding no filaments "sufficiently satisfactory, when looked at in a commercial sense." William Sawyer's "small sun" suddenly seemed America's best shot, but Sawyer was no Edison. A notorious drunk given to violent rages, he alienated engineers and investors alike. Raging at colleagues, filing lawsuits, later shooting a neighbor, Sawyer died in 1883, a victim of his own temperament and an insatiable taste for brandy.

Another inventor chasing incandescence was Hiram Maxim. Working in his Brooklyn laboratory, this burly native of Maine had already earned patents for the basic mousetrap, a curling iron, and in 1878, for "Improvement in Electric Lamps." Maxim's bulb, his patent noted, "shall be compact and of small size, so as to cause very little shadow, of slight cost in construction, susceptible of very delicate adjustment, and suitable for use where great nicety and steadiness are required." Maxim would eventually earn eighteen patents for lighting, and would worry Edison by installing the first electric lighting in a Manhattan office. But like many inventors, Maxim was a one-man idea factory who thought small—of single homes and compact systems. He saw "too many unsolved technical problems" to imagine electricity in all homes. Within a few years, frustrated by Edison's spreading acclaim and under suspicion of bigamy, Maxim would move to England. There he would invent a multiwing flying machine that never flew, several engines, and the device for which he was knighted—the machine gun.

In the fall of 1878, Edison rejoined the quest. "It was all before me," he later said. "I saw the thing had not gone so far but that I had a chance. I saw that what had been done had never been made practically useful. The intense light had not been subdivided so that it could be brought into private houses." A month after resuming his search, Edison added a thermal regulator to control heat and had a working model to show the press. "There was the light," the *New York Sun* wrote, "clear, cold, and beautiful. The intense brightness was gone. There was nothing irritating to the eye. The mechanism was so simple and perfect that it explained itself." The *Sun* reporter did not note—because Edison did not tell him—that the bulb lasted only an hour. Instead, with backing from Wall Street, the Edison Electric Light Company issued its first stock.

No solitary genius toiling in a dark chamber, Edison had gathered the East Coast's best engineers, whom he called his "muckers," in his laboratory in Menlo Park, New Jersey. From this greatest of all his inventions, the research and development lab, Edison promised "a minor invention every ten days and a big thing every six months or so." But such a pace would have been impossible without Edison himself. Though nowadays more a legend than a man, he was then just thirty-one, possessed of enormous confidence, shrewd commercial sense, and all that perspiration he would later tout as the essence of genius. In the fall of 1878, news of Edison's little light sent gas company stocks plummeting, but his boast of someday lighting "the entire lower part of New York City using a 500 horse power engine" still hung on his search for a lasting filament.

Electric light now flows from many sources—the mercury vapor of fluorescents, the ionization of neon lighting, the electron discharge of light-emitting diodes (LEDs)—but in 1878, the only source was resistance. Even inventors confused by Maxwell's equations knew that a filament hooked to a battery brightened as its atoms resisted the current. Edison stated the equation: "The more resistance your lamp offers to the passage of the current, the more light you can obtain with a given current." But resistance also creates heat that can snuff out its own light. Edison tried filament after filament, keeping careful notes. Silicon. Cork. Boron. Rosewood. In his notebooks, failures were followed by "T.A.", short for "try again." Hickory—T.A. Coconut—T.A. Cotton soaked in tar—T.A. Fishing line . . . "I have carbonized and used cotton and linen thread," Edison wrote, "wood splints, papers coiled in various ways, also lamp black, plumbago, and carbon in various forms, mixed with tar and rolled out into wires of various lengths and diameters." He had even tried hair from the beards of two machinists. His men bet which hair would burn out first. Neither lasted long.

By the summer of 1879, Maxwell's last and Einstein's first summer, there was growing speculation that before the end of the year some wizard would flip a switch and a small, smokeless bulb would glow, not just for a moment but for hours. The once-skeptical *New York Times* now envisioned "cities and villages lighted by the electric energy . . . farm houses running their own little generators . . . gas and oil being remembered as barbarians of the past." All summer and on into the fall, the competition continued. William Sawyer drank. Hiram Maxim took ill. Thomas Edison and his muckers worked on.

Finally . . .

On October 22, 1879, Edison was burning the midnight oil in a pursuit that would soon make that phrase just a figure of speech. That evening, his assistant Charles Batchelor noted "some very interesting experiments on straight carbons made from cotton thread." At one thirty A.M., Edison and Batchelor connected their latest model to a battery. The little bulb glowed softly, burning on into the night. When the sun rose, the bulb was still shining. Batchelor guessed it to be as bright as thirty candles, or four kerosene lamps. And on it burned, throughout the morning and into the afternoon. Finally after fourteen hours, Batchelor upped the voltage. The filament flared; the bulb cracked. Its vacuum broken, the glass globe filled with air, oxygen flaring the filament and snuffing the light. Using a carbonized cotton thread, Edison's breakthrough bulb resembled the designs of William Sawyer, Joseph Swan, and Hiram Maxim, but Edison soon showed how *he* was altogether different.

Edison already had 250 patents, but on November 4, he filed for another, then fired up his assistants with a deadline. The Menlo Park machine shops began mass-producing filaments, glass, heat regulators, wires, switches, fuses, and other apparatus. Edison planned to invite the press and public into his lab on New Year's Eve. A week before Christmas, he let the *New York Herald* offer advance publicity.

ELECTRIC ILLUMINATION
—A SCRAP OF PAPER—
IT MAKES A LIGHT, WITHOUT GAS OR FLAME,
CHEAPER THAN OIL—
SUCCESS IN A COTTON THREAD

The *Herald* reporter could hardly contain his amazement. "Edison's electric light, incredible as it may appear, is produced from a tiny strip of paper that a breath would blow away. Through this little strip of paper passes an electric current, and the result is a bright, beautiful light, like the mellow sunset of an Italian autumn." Reading of Edison's New Year's Eve display, crowds began arriving in Menlo Park. Extra trains added to the line brought people from surrounding states. The light met them at the train station: twenty glowing street lamps leading to Edison's compound on Christie Street. As promised, when the sun set on the last night of the year, Edison opened his lab. Hundreds filed into the covered sheds to see the modern miracle. Twenty-five bulbs lit one lab, eight more shone in Edison's office. While his assistants explained how efficient electric lighting would soon become, Edison turned the lights on and off—and on again. Farmers and bankers, women in furs, men in top hats, all stood admiring the wizard—not the first, yet the most persistent, the most savvy—standing beside what one writer called "a little globe of sunshine."

Over the next decade, incandescent lighting would cause battles that made Particle vs. Wave look like a skirmish. Electric light had arrived, but how would it be delivered and sold? Even as Edison's lights glowed in Menlo Park, arc lamps flickered in hotels, mansions, and on a few streets. One summer evening in 1880, several months after Edison's New Year's Eve party, Ohio farm boy Charles Brush waited for darkness, then threw a switch to start his generator. The streets of Wabash, Indiana, were lit, as a reporter wrote, by a "strange weird light, exceeded in power only by the sun." Just a single cluster of arc lamps near the Wabash town hall had lit the entire downtown. "The people, almost with bated breath, stood overwhelmed with awe, as if in the presence of the supernatural." Night trains passing Wabash began slowing so passengers could admire the aura. Arc lamps soon flared in Denver, Minneapolis, and San Jose, California, yet once initial awe abated, many found the light harsh and eerie. When Brush's arc lamps lit up Broadway, leading to its nickname, "The Great White Way," some carried umbrellas to block the glare. Lighting whole squares with a few globes was also impractical. Arc lamps, perched atop towers, were still too bright to be looked at. Birds awoke and sang. People squinted. Edison, again, had an answer.

Not long after his New Year's Eve show, Edison sent assistants to exotic places in search of a longer-lasting fiber. He tested some six thousand more substances before settling on bamboo. Meanwhile his Menlo Park machine shop churned out more bulbs, more switches, more wires and fuses. And on September 4, 1882, after a summer of frenzied labor and expenditures exceeding half a million dollars, Edison, in a dark frock coat and a white derby hat, stood in the Wall Street office of Drexel-Morgan. As three P.M. approached he checked his watch. A few blocks away, at 257 Pearl Street, Charles Batchelor checked his. When the hour arrived, Batchelor turned on an enormous generator. With a great groan, and sparks that terrified some mechanics, a current began flowing through copper wires buried beneath the carriage-clogged streets of Manhattan's Financial District. On his end, Edison was poised at the switch. A shout came from the back of the office.

"One hundred dollars they don't go on!"

"Taken!" Edison said.

A moment later, he closed the switch beside him and throughout Lower Manhattan, four hundred little globes of sunshine sparkled. Edison accepted congratulations, but it took sunset to bestow a full appreciation. Edison had made sure the *New York Times* offices were among those lit. Now the *Times* rewarded his PR skills:

> It was not until about 7 o'clock, when it began to grow dark, that the electric light really made itself known and showed how bright and steady it is . . . It was a light that a man could sit down under and write for hours . . . The light was soft and mellow and grateful to the eye and it seemed almost like writing by daylight, to have a light without a particle of flicker and with scarcely any heat to make the head ache.

Another struggle awaited, the epic "war of the currents." Edison favored low-voltage direct current (DC) requiring generators just a few miles apart. DC, he said, was "all I'll ever fool with." But the brilliant Serb emigré Nicola Tesla promoted high-voltage alternating current (AC), which could be brought into every home at vast distances from the power source. Edison found AC's three thousand volts terrifying. When the industrialist George Westinghouse backed Tesla, Edison predicted, "Just as certain as death Westinghouse will kill a customer within six months after he puts in a system of any size." No customers died, but several workmen who touched live wires did, as did a circus elephant whose public electrocution Edison arranged to alert the public to the danger. But a new invention, the transformer, lowered high voltage for home use. After still more patents, and hundreds of lawsuits, AC became the standard current. By then, as if playing out the prophecies of Zoroaster and Mani, light was battling darkness in great orgies of incandescence.

- 1887: Thousands of colored lights beam on a 150-foot fountain soaring from the shores of Staten Island.
- 1889: Paris's Exposition Universelle, marking the hundredth anniversary of the French Revolution, glows with electric lights totaling 176,000 candlepower, including a fully lit Eiffel Tower with a beacon at its crown.
- 1893: Chicago's "White City" blazes with 200,000 lights.
- 1894: Strings of lights line the fairy-tale domes and towers of Steeplechase Park in Coney Island.
- 1901: Buffalo's Pan-American Exhibition is "a city of Living Light," where 40,000 bulbs outline a single building and a statue honors the Goddess of Light.

An early ad for Edison's light listed its advantages: "It is the most economical artificial light; It is brighter than gas; It is steady as sunlight; never flickers . . ." Yet the allure of electric light transcended cost and convenience. Anyone could make candles and lamps, and anyone could understand them—they burned. But electricity was as ethereal as the sun itself, and its light bred the same fascination that surrounds today's digital devices. "Any sufficiently advanced technology is indistinguishable from magic," the science fiction writer Arthur C. Clarke famously noted. Adding to light's magic was human ingenuity. No one knew who had made the first lamp, but the men who turned amber's fleeting sparks into pure, reliable light became household names. Edison. Tesla. Westinghouse. The light they made seemed to have few limits, even if each bulb lasted only a few weeks. With the flick of a switch, light—long a force by day and glimmer by night—was coming, house by house, to simplify domestic life. All too soon it would be taken for granted, yet in these first decades after Faraday, after Maxwell, after Edison, light bore the promise of a new century.

CHAPTER 13

c: *Einstein and the Quanta, Particle, and Wave*

And pluck till time and times are done
The silver apples of the moon
The golden apples of the sun.

—WILLIAM BUTLER YEATS, "THE SONG OF WANDERING AENGUS"

Throughout the autumn of his seventeenth year, in an era gripped by inertia, Albert Einstein was busy becoming Albert Einstein. That fall of 1895, England's Queen Victoria was approaching her sixtieth year on the throne. Aged dynasties—the Hapsburgs, the Romanovs, the Ottoman Empire—stood solidly on their pedestals. Art and science were evolving, but physics had stalled. "The more important fundamental laws and facts of physical science have all been discovered," one professor told a class, "and . . . the possibility of their ever being supplanted in consequence of new discoveries is exceedingly remote."

But the man who would become Einstein was just beginning to bloom that autumn. Having failed the entrance exam at Zurich Polytechnic—he aced physics but knew little of French or botany—he was sent to a boarding school west of Zurich. There he lived with a family of seven children whose father shared Einstein's contempt for all things rigid or conservative. At school, fellow students considered this German transplant a "laughing philosopher," yet he was more serious than they imagined. At sixteen Einstein had renounced his German citizenship, rejected Judaism, and begun "a positively fanatic orgy of free thinking." Bored in class, he took long walks in the mountains, experimenting in his own private chamber—his mind.

All his life, he would perform gedankenexperimente—thought experiments. That fall, he did his first. What would the world look like, he wondered, if he could ride on a beam of light? "If a person could run after a light wave with the same speed of light," he thought, "you would have a wave arrangement which could be completely independent of time." He knew that "such a thing is impossible," yet he would spend the next decade pondering the ride. Viewed from a beam of light, clocks would seem frozen, the light from their moving hands never reaching the retreating passenger. And yet, like someone in a slow elevator, light's rider would barely know he was moving. All the staid laws of Newton's universe would be upset.

Early in the twentieth century, light, so long ephemeral and elusive, became the anchor of the cosmos. First its speed (in a vacuum) was shown to be constant: 186,282 miles per second. Coming at you. Going away. In a lab, beamed from a train or plane, across galaxies, or in the mind of Albert Einstein—186,282 miles per second. As this startling truth settled in, new challenges dizzied the mind. Time slowed. Space curved in on itself. Then the old argument—Particle vs. Wave—returned. The nagging possibility that light might somehow be both particle *and* wave raised doubts about what, if anything, was certain. There remained but one fixture, unchanging and unsurpassable—the speed of light. In equations, the speed was a single letter, *c*, for the Latin *celeritas*, meaning swiftness. As the century unfolded, this constant dared, as T. S. Eliot would have it, to "disturb the universe."

First, however, light's unfathomable speed had to be measured.

Light's earliest students had wondered whether light even had a speed. Empedocles thought so, but Aristotle argued that light was "not a movement." Hero of Alexandria, the designer of the first steam engine, put light's speed to a test. Step outside at night, Hero wrote, and look skyward. Close your eyes. Open them. Your instant perception of distant stars proves that light "is emitted with infinite speed." The argument continued, Islamic scientists considering the speed finite, Kepler and Descartes arguing that it was instantaneous. Then, in 1676, a Dutch astronomer working in Paris used the clockwork of the planets to finally time a beam of light.

In the half century since Galileo had first seen Jupiter's moons, their pinpoints of light had been charted, each pass behind the giant planet noted down to the second. Comparing these timetables to his own observations, Ole Roemer spotted a discrepancy. When the Earth and Jupiter were on opposite sides of the solar system, the Jovian moon Io emerged several minutes behind schedule. In August 1676, Roemer made a bold prediction: on November 9, with Jupiter at its farthest from Earth, Io would not reappear at five twenty-seven P.M. Instead, its light, delayed by travel across the solar system, would arrive eleven minutes late. When Roemer's prediction proved precise, Christiaan Huygens triangulated the distance to Jupiter, divided distance by time, and announced a speed of light that astounded even Isaac Newton—144,000 miles per *second*. Though nearly 25 percent too slow, it was a start.

A half century later, a British astronomer named James Bradley used "stellar aberration," the apparent shift of the stars due to the Earth's orbital velocity around the sun, to come closer to light's exact speed. Sunlight, Bradley announced, reaches the Earth in eight minutes and twelve seconds. He was eight seconds off. Then in the mid-1800s, as the rest of Paris posed for daguerreotypes, two French physicists gave light a more precise speed. In 1849, Hippolyte Fizeau sent a beam of limelight, made by burning quicklime, through a spinning cogwheel. Chopped into pulses, the beam sped to the hills of Montmartre and back. Timing the pulses, the shaggy-bearded Fizeau measured the speed of light at 196,476 miles per second. Closer, closer . . . Thirteen years after Fizeau's test, Léon Foucault, already famous for his pendulum proving that the Earth rotates, fired a beam at a mirror whirling at eight hundred rpm, then bounced it to a stationary mirror twenty meters away. By the time the beam returned, even at the speed of light, the spinning mirror had moved a fraction of a degree. Measuring the angle defined by the beam coming and going, calculating the mirror's fraction of a revolution, Foucault clocked light at 185,200 miles per second. Foucault's measure lasted until the late 1870s, when a young American ensign turned his attention to light.

Albert Michelson, son of Prussian immigrants, grew up in the mining camps of the Sierra Nevada during the Gold Rush. Like many landlocked boys, Michelson dreamed of the sea. Too young to serve in the Civil War, he sought an appointment to the U.S. Naval Academy but was turned down. Short and wiry, confident and driven, Michelson crossed the continent by foot, horseback, and the new Transcontinental Railroad to reach the White House. There he schemed to meet President Ulysses S. Grant, convinced him of his abilities, and became an Annapolis cadet. As an ensign, Michelson measured his ship's speed by taking readings of wind and water. Something in the process of calculating against the wind, across the water, stayed with him. A few years later, while teaching physics at the Naval Academy, Michelson read of Foucault's results for the speed of light and thought he could come closer still. The quest for a precise speed of light would consume him for the rest of his life.

Michelson's equipment—a lens, a steam boiler, a tuning fork, and two mirrors, one fixed, one spinning—cost him ten dollars. For such a price, he would measure light with the precision of Newton and his prisms, Fresnel and his integrals, Maxwell and light itself. Each of Michelson's many tests began an hour before sunrise or sunset, when he found light "sufficiently quiet to get a distinct image." In a storage shed perched on the seawall of Annapolis's Severn River, Michelson fired up the boiler. Within minutes, its steam spun the mirror, slowly at first, then so fast it sometimes flew off its mooring. When the mirror, timed to match the vibrations of the tuning fork, was whirling at 257 revolutions per second—yes, per *second*, in 1879—Michelson was ready. Aiming his lens at the horizon, he caught sunlight and bounced it off the whirling mirror. The beam split the leafy campus. Striking a fixed mirror precisely 1985.09 feet away, the beam scorched back to the glass that had since spun a fraction of a fraction of a revolution. The exacting Michelson filled his log with data—"Date of Test," "Distinctness of Image," "Temperature," "Speed of Mirror," "Displacement of Image" . . . but only one number really mattered. It varied by a few kilometers with each test, but on average, Albert Michelson's measurement of light—186,319 miles per second—just 0.0002 percent above the actual speed.

Michelson claimed he studied light "because it's so much fun," but he lived in awe of it. In lectures, he urged budding physicists to notice "the exquisite gradations of light and shade, and the intricate wonders of symmetrical forms and combinations of forms which are encountered at every turn." Yet aesthetics never interfered with his determination to pinpoint light's speed. During the 1920s, when he was in his seventies, Michelson would beam light across mountaintops in Southern California, fixing a speed—186,285 miles per second—that would stand until the age of lasers. But as a dapper young officer, having made headlines with his measurement, he turned to the most enduring of light's enigmas—the ether.

Abhorring the idea of a vacuous universe, Aristotle had filled it with aether, named for the Greek god of light or upper air. Ever since, "luminiferous ether" had been a fixture of the cosmos. No one had seen ether, yet all knew it well. The Roman poet Lucretius wrote of "the aether the stars graze upon." Newton considered it "solary fewell," a substance like air yet "rarer, subtiler & more strongly Elastic." James Clerk Maxwell described ether in the *Encyclopedia Britannica*: "There can be no doubt that the interplanetary and interstellar spaces are not empty, but are occupied by a material substance or body, which is certainly the largest, and probably the most uniform body of which we have any knowledge."

For all but a few skeptics—Michael Faraday among them—light was unimaginable without ether. The parallel was obvious: just as sound rippled the air, light disturbed the ether. To consider that ether might someday prove as whimsical as phlogiston, an imaginary substance once thought to transmit heat, was to doubt sun and moon. By the 1880s, while closing in on light's speed, physicists were seeking another speed, that of the "ether wind."

Toward the end of his frail life, Augustin-Jean Fresnel had suggested that the Earth passing through the ether would create an "ether drag." Such friction, however small, would slow light's blazing speed. Fresnel calculated an "ether drag coefficient," yet not even his exactitude could detect such a minuscule effect. Then in 1849, shortly after timing the speed of light across Paris, Hippolyte Fizeau tested Fresnel's theory. If parallel beams were sent through tubes of rushing water, light should go faster with the flow, slower against it. Fizeau tested and retested, but regardless of the water's direction, the change in speed was insignificant. Still, faith in the ether persisted. More precise instruments would surely detect it. Because the Earth traveled through space at thirty kilometers (18.6 miles) per second, "ether drag" should slow or speed light by that much, depending on the direction of the beam measured. Thirty kilometers per second, however, is one ten thousandth the speed of light, and measuring such a minuscule time would require unprecedented precision. By 1881, Albert Michelson had all the tools, mental and physical.

While studying in Berlin, Michelson devised an ether speedometer. An upgrade of his invention, now called an interferometer, can be found in every modern optics lab. Michelson's device used a thinly silvered mirror that let some light pass but reflected the rest. Hitting the mirror at 45 degrees, a single beam became two—one passing through the mirror and a second reflecting at a right angle to the first. Each bounced off a mirror on the rim of the device and returned to the source, forming a + of light. Fresnel's "ether drag" should slow the velocity of one beam but not the other. Explaining the experiment to his children, Michelson imagined swimmers racing in a river, one swimming across a current, the other downstream: "The second swimmer will always win, if there is any current in the river." As his perpendicular beams overlapped on their return to the source, Michelson expected to see them out of sync, the clue being the interference stripes Thomas Young had first seen. But peering into his lens, Michelson saw no dark lines, only pure light. There was no current in the river. After several tests, Michelson wrote to his financial backer, Alexander Graham Bell.

> My Dear Mr. Bell,
> The experiments concerning the relative motion of the Earth with respect to the ether have just been brought to a successful termination. The result was, however, negative.

Unseen, undetected, ether was not so easily dismissed. Soon it was even appearing on America's Chautauqua circuit, which brought professors and pundits to rural towns and villages. Standing before rustic crowds in overflow tents, one physicist held up a plate of jelly. "One thing we are sure of," he said, "and that is the reality and substantiality of the luminiferous ether . . . an elastic solid, for which the nearest analogy I can give you is this jelly."

Ether just *had* to exist. Light could not propagate through an empty universe. Another test was needed, and in 1885, Albert Michelson was ready to try again. This time he had a partner—Edward Morley. A former preacher turned chemist with the buttoned-down propriety of a college dean, Morley soon grew worried about Michelson. Obsessed with light's speed, the diminutive physicist had begun skipping meals and sleep, driving himself, as Morley noted, "to a task he felt must be done with such perfection that it could never again be called into question." The breakdown was not long in coming. In September 1885, Michelson suffered a nervous collapse and was sent to a sanitarium. Morley, suspecting "softening of the brain," predicted his partner would never work again, yet by the following spring, both men were back in the lab.

To measure even the smallest ether wind, Michelson built a steel X with twelve-foot arms embedded in a sandstone slab. Next he added mirrors, sixteen in all, that would bounce the light back and forth, back and forth. Each mirror was fine-tuned by a screw laced with a thousand threads per inch. Finally, Michelson floated the entire device on a pool of mercury, allowing it to slowly spin as light swam upstream and down at all angles.

July 8, 1887. Noon. A basement at the Case School of Applied Science in Cleveland, Ohio. The stage is set for the "most famous failed experiment" in the history of science. The balding, bespectacled Morley sits in one corner while Michelson, now sporting a handlebar mustache, circles the giant slab, calling out numbers. Yellow light, from salt burned in an Argand lamp, bounces off the sixteen mirrors to weave a lattice of beams. Michelson calls out more numbers; Morley scribbles. And as they measure the light zipping against and with the fabled ether wind, they detect not the slightest change in speed. "If there be any relative motion between the Earth and the luminiferous ether," Michelson will report, "it must be quite small, quite small enough entirely to refute Fresnel's explanation of aberration."

For the next two decades, physicists lived in denial. "Remove from the world the luminiferous ether," wrote Heinrich Hertz, whose discovery of radio waves proved Maxwell's electromagnetic theory, "and electric and magnetic actions can no longer traverse space." Further tests of ether wind used longer light paths, whirling discs, the thin air of mountain labs, and the swirling air in gondolas hung from balloons. The tests only proved Einstein's later definition of insanity—doing the same thing over and over and expecting different results. At the dawn of the twentieth century, luminiferous ether was still being touted in textbooks with only the slightest suggestion that it might not exist. Albert Einstein, however, was still riding a beam of light.

"If I pursue a beam of light with the velocity *c*," he imagined, "I should observe such a beam of light as an electromagnetic field at rest though spatially oscillating . . ."

Einstein had spent only a year in the small Swiss canton where he conceived this thought experiment, yet its enigma continued to cause a "psychic tension." The tension followed him to Zurich Polytechnic where, for his senior thesis, he proposed studying light's speed through the ether. Learning that this had been done, he read about the Michelson-Morley experiment. Unlike those eager to deny its conclusion, Einstein accepted the results. Having already cast off his country and his religion, he had little trouble rejecting the existence of ether. If what he called "the light medium" did not exist, Einstein reasoned, then light must travel at the same speed at all times, in all directions. This made his mental beam still more perplexing. How could two observers, one traveling, one standing still, perceive the same speed of a beam? "The difficulty to be overcome," Einstein later wrote, "then lay in the constancy of the velocity of light in a vacuum, which I first thought would have to be abandoned."

Einstein finished his Ph.D. thesis in 1901 but his professors took their time with accepting it. The following year, so desperate for work that he considered playing his violin on the street, he took a job as a "patent slave," reviewing patents in the Swiss capital of Bern. He and his new wife, Mileva, lived on the Kramgasse, just down the street from Bern's medieval clock tower. The clock featured wheels within wheels to show the movements of the heavens and a mechanical jester in cap and bells whose hammer chimed each hour. Einstein strode past the tower each morning to reach his third-floor office, where he summarily reviewed the day's patents, leaving him afternoons to clutter his desk with his own work. Among his inquiries was the latest of light's surprises.

In 1902, the same year Einstein became a "patent slave," a German physicist discovered that when ultraviolet light strikes a metal plate, it knocks off electrons. Given that light is electromagnetic, this was not strange, yet there was a twist. Logic suggested that stronger light should release electrons with higher kinetic energy, yet no matter how bright the light, all the freed electrons had the same energy. The only way to boost the energy was to use light of a different color. Tightly wound blue waves freed electrons, but longer red waves did not. But if made of waves, how could light knock off individual particles? Another few years passed at the patent office. The jester on Bern's clock tower chimed each hour. Then came 1905.

Scholars still struggle to explain Einstein's emergence. He later recalled that "a storm broke loose in my mind," yet Einstein's annus mirabilis, as vital to the understanding of light as Newton's year of prisms, had several sparks. Einstein had been reading physicist Ernst Mach: "No one is allowed to predicate things about absolute space and absolute motion. They are pure things of thought, pure mental constructs." Einstein also knew the theories of the French mathematician Henri Poincaré, who had proposed the relativity of time, yet clung to his belief in the ether. Equally important was Einstein's independence. As a patent clerk he was under no pressure to churn out papers; as an expatriate German and nonpracticing Jew he was free of cultural constraints. And then, of course, he was Einstein. Early in 1905, he set aside the "psychic tension" of riding on a beam to consider light's release of electrons. That March he submitted the first of his Wonder Year's four papers. He considered it "very revolutionary."

Einstein's revolution faced down one of physics' pressing problems: How does matter emit heat? Consider a chunk of iron in a fire. As it heats, the iron absorbs the fire's energy and emits some. Physicists call such heat blackbody radiation. As it intensifies, blackbody radiation becomes visible once the spread of electromagnetic wavelengths reaches from the infrared into the red part of the spectrum. Keep heating the iron and the emission will turn yellow, then blue, and even trespass into the ultraviolet. When white-hot, the iron will emit an amalgam of all the colors of the visible spectrum. Allowed to cool, the iron loses energy, its color changing accordingly. But what if it were not allowed to cool? Imagine, instead of iron, a box of mirrors with only a small opening. Pour heat into it, allowing little to escape. As the box heats, wouldn't the energy build up, changing its wavelength of peak emission from red to yellow to blue to ultraviolet? And wouldn't the box absorb so much energy that it exploded? This paradox, confounding physicists by defying their time-tested theories, became known as "the ultraviolet catastrophe." Its solution would change everything. Everything.

One of many perplexed by the "catastrophe" was Max Planck. Mournful, mustachioed, his blank face a perpetual riddle, the German physicist was a conservative family man proud of all things Teutonic, including the Bach and Brahms he played on the piano. Planck had no intention of starting an upheaval in physics. He merely sought an answer to the "ultraviolet catastrophe." Planck spent years heating iron boxes and calculating the emissions from a small hole in each. Instead of rising and rising, however, the radiation peaked at a particular wavelength, then descended in a predictable curve. Some property of energy was essentially venting the buildup. "For six years I had struggled with the blackbody theory," Planck recalled. "I had to find a theoretical explanation at any cost." Finally, in "an act of desperation," Planck devised a different answer.

Perhaps energy came not in waves but in packets. Emissions from such packets would not be graphed as a steady curve but as a staircase, its steps rising as each new energy packet was emitted. Planck charted the emissions, calculated the steps, and saw each jump at a steady rate now called Planck's constant. The constant was incredibly small—a decimal preceded by two dozen zeros—yet when Planck applied it to blackbody radiation, the numbers added up. Higher frequencies of light, blue or ultraviolet, must come in larger energy packets. A blackbody would absorb abundant heat at low frequencies but when intensified into higher frequencies, the packets would be larger and fewer. It was as if the crowd at an open-air concert, instead of arriving on foot and overflowing the field, had come only in cars and buses whose numbers were limited by parking. Packets kept attendance down, keeping a lid on things. This seemed barely plausible, but Planck's constant suggested more startling scenarios.

Leaping from step to step, Planck's energy packets defied the laws of classical physics discovered by Galileo and Newton. A packet could contain x amount of energy, or $2x$ or $3x$ . . . but nothing in between. Imagine a car going ten miles per hour, then instantly going twenty, without ever going eleven, twelve, thirteen . . . Imagine an elevator that did not rise from floor to floor but suddenly *appeared* on the second floor without passing between first and second. Planck's theory seemed absurd, but mathematically it dodged the "ultraviolet catastrophe."

Planck named his energy packet a *quantum* (plural *quanta*; from Latin *quantus*, "how much?"). The sad-eyed physicist was not convinced that quanta existed; he only knew that he had solved the riddle. Planck announced his theory on December 14, 1900. The date gave the twentieth century, commonly acknowledged to begin in 1901, a two-week head start. It would not be the last time that quanta toyed with time. Most physicists ignored Planck's quanta, but Einstein recalled, "It was as if the ground had been pulled out from under one." By 1905, he was building a new foundation.

Considering "the photoelectric effect"—how light blasts electrons off a metal plate—Einstein's "very revolutionary" paper used quanta. "According to the assumption to be considered here," Einstein wrote, "when a light ray is propagated from a point, the energy is not continuously distributed over an increasing space but consists of a finite number of energy quanta which are localized at points in space and which can be produced and absorbed only as complete units." Only if light was quanta—*Lichtquanten* in Einstein's German—could it explain the photoelectric effect. The patent clerk used Planck's constant to calculate energy absorption and emission. As to why differently colored light released electrons of a different kinetic energy, Einstein noted the amounts of energy in each hue. "The simplest conception is that a light quantum transfers its entire energy to a single electron."

The existence of Einstein's *Lichtquanten* would not be proven for another decade. Max Planck objected, saying that quanta did not apply to visible light, which everyone, even Einstein, knew to be a wave. Einstein's explanation of the photoelectric effect included a caveat: "I insist on the provisional character of this concept which does not seem reconcilable with . . . the wave theory." But Einstein was too busy to ponder particles and waves. Following the publication of two more papers in the spring of 1905, he was again riding a light beam and wondering.

How could light travel at the same speed regardless of its direction or source? Such a constant speed defies common sense. When I walk on an airport's moving sidewalk, my speed is added to its speed. My legs carry me three miles per hour, the sidewalk moves at five, and my combined speed is eight miles per hour. So I get to wait for my luggage a minute sooner. But as the Michelson-Morley experiment showed, light defies this simple summation. Shoot a beam of light from a speeding jet, aim it forward, backward, sideways, and it *always* travels at the same speed. The same damned 186,282 miles per second. Light simply refuses to behave like anything else in the universe. Its constancy seems impossible, unless . . .

Late one morning in May 1905, Einstein was walking through Bern with his fellow patent clerk, Michele Besso. Einstein often shared his latest thought experiment with Besso. Now he confessed despair over light's constancy. "I'm going to give it up," Einstein said. Then, right there on the Kramgasse, as trams passed and pedestrians jostled, the answer came to him. From Poincaré, from Mach, from his own unfathomable genius, he *saw.* If light traveled at a constant speed, as seen by all observers, on all galaxies, in all tests, then time must be mutable. The next day, Einstein rushed up to his friend. "Thank you," Einstein told Besso. "I've completely solved the problem."

Einstein spent five weeks writing "On the Electrodynamics of Moving Bodies." He would spend the rest of his life explaining it, sometimes with equations, sometimes with a joke: "Put your hand on a hot stove for a minute and it seems like an hour. Sit with a pretty girl for an hour and it seems like a minute. That's relativity." Nowhere in his paper did Einstein suggest riding a beam of light, yet he later called the idea the "germ of the theory of special relativity." The constant speed of light, Einstein saw, made time "relative."

Picture two observers, one riding a train, the other sitting on an embankment watching the train pass. The passenger puts a flashlight on the ceiling of the coach and a mirror on the floor. Turning the flashlight on and off, he sees a beam travel straight down and bounce straight up, its path forming a vertical line. Now imagine what the observer on the embankment sees. From this point of view, the light beam is carried horizontally by the moving train. It hits the floor at an angle and bounces back at a similar angle. The shape is not an I but a V. Clearly light's V-shaped path is longer than the ricocheted I. Yet if light's speed is constant, how can it travel different distances in *exactly the same time*? Time is "the culprit," Einstein realized. If light is constant, time cannot be. The only way to explain the paradox was to believe—begin to believe—that time as perceived by a stationary observer ticks more slowly on moving bodies.

It sounded—still sounds—incredible, ridiculous, even, but Einstein used an equation, the Lorentz transformation, to calculate how time slows negligibly at jet speed but nearly stops as an object approaches the speed of light. (If the train were approaching 186,282 miles per second, the V seen from the embankment would be enormously long, while the I observed from the train would be the same. Light would trace each letter at its constant speed, forcing time, as experienced by the passenger, to nearly stop.) And time has since proved Einstein correct. Seven decades after Einstein conceived relativity, atomic clocks had become sufficiently precise to test time dilation. A pair of clocks were flown around the world, then compared to stationary clocks. The moving timepieces returned a fraction of a second slower than the stationary clocks, precisely as calculated by Einstein. And when today's particle accelerators blast subatomic muons from an atom, each survives for microfractions of a second. But when accelerated to approach the speed of light, their clock slows and the particles survive slightly longer. Time is the culprit, light its accomplice.

Einstein's theory of special relativity met with more bewilderment than acceptance. The public heard nothing of it, but a few physicists, including Max Planck, found it intriguing. After four more years in the patent office, Einstein was given a professorship in Berlin and had more time for thought experiments. Soon these experiments focused on gravity. Michael Faraday had used an enormous magnet to twist polarized light; Einstein would use the sun to bend it.

By the summer of 1912, faith in luminiferous ether was waning. The Michelson-Morley experiment had become widely known. Einstein had judged "the light medium" to be "superfluous," and aside from a few diehards still running experiments, the ether that never existed was fading into obscurity. Thomas Edison's light glowed in city hotels and upper-class homes, Daguerre's light was being captured by millions of Kodak cameras, and the Lumière Brothers' moving light was fashioning an industry of illusion. Yet the preponderance of light on Earth still streamed from the heavens. Thus it was the sun, together with its planets, that inspired the first accurate model of how light is actually made.

That summer, the young Danish physicist Niels Bohr took time out from his honeymoon to tinker with the atom. Although born in the century of Faraday and Maxwell, Bohr was a thoroughly modern man. He loved modern art, especially cubism. "It is for the painters to find something new," he often said. Profound and pontifical, the jowly Bohr always spoke slowly, lighting his pipe, punctuating his utterances with silence. Einstein likened Bohr to "an extremely sensitive child who moves around the world in a sort of trance." But Bohr had no childlike awe of light. As a theoretical physicist he worked not with mirrors and lenses but with pencil and paper. Throughout the summer and fall of 1912, researching in England, he sought a workable model for the atom. His boss, Ernest Rutherford, had proposed a tiny nucleus—"like a fly in a cathedral"—circled by electrons in a single orbit. But wouldn't orbiting electrons lose energy and crash into the nucleus? "It was clear," Bohr said of Rutherford's atom, "[that] we could not proceed at all in any other way than by radical change." After seven months, Bohr realized the answer had to involve quanta.

Instead of one orbit around each nucleus, Bohr suggested many—an atom-sized solar system. When stable, an electron orbits the nucleus at an inner "stationary state," but when hit by energy—light, heat, electricity—the electron absorbs its packets. The excited electron leaps from an inner to an outer orbit. If no more kick is added, the electron falls back to its inner orbit, releasing the absorbed energy as a single quantum of light. Bohr's atom explained the mysterious colored stripes emitted by burning chemicals in telltale patterns that acted as fingerprints, letting astronomers learn the composition of the stars. Many were skeptical of Bohr's atom, yet Einstein called it "an enormous achievement." Bohr's quantum leap, along with updating the atomic model, paved the way for neon lights, fluorescents, and LEDs. Explaining the light of sun and stars, however, would take subtler thinking.

A year after Bohr quantized the atom, Einstein got a divorce. Mileva took their two sons, leaving Einstein alone. Anonymous outside the scientific community, holed up in a disheveled apartment in Berlin, he had time to think, to imagine, to ride again.

In this thought experiment he imagines riding an elevator, a cubicle steadily accelerating toward the stratosphere. The motion keeps him pressed to the floor, going faster, faster. But without windows, unable to sense his motion, he cannot be sure whether his sense of "weight" is due to gravity or acceleration. He reasons the two to be, in effect, the same. This suggests numerous possibilities, the most startling being the bending of light. A beam shot through a hole in the rising elevator would strike the opposite wall slightly lower than where it entered. Acceleration would appear to bend light's rigid beam. And because acceleration is, to the passenger, no different from gravity, Einstein wondered whether gravity might also bend light. With another storm loose in his brain, he wrestled with vectors and tensors (measuring movement across curved space), then churned out pages of calculus, non-Euclidean geometry, and other "gravitational equations." When it was finished, in 1915, Einstein regarded his Theory of General Relativity as a work "of incomparable beauty." The theory also forced light to further disturb the universe.

Gravity, Einstein argued, is not a force, as Newton conceived it. Instead, like a bowling ball resting on a blanket, matter curves the space around it. Stars, moons, and planets curve surrounding space, attracting nearby objects and, Einstein proposed, bending light. Just like the beam crossing his rising elevator, starlight passing the sun should veer. Such a curvature should be observable during a solar eclipse. Einstein hoped such a test could be made soon, but eclipses came and went, foiled by clouds or the Great War. Then, in May 1919, two British teams set out, one for the Amazon, the other for West Africa, to photograph an eclipse and track the trajectory of starlight passing the blackened solar disc. Six months later, with the plates developed, and the results confirmed, an overflow crowd filled an auditorium in London. The scene, one observer noted, resembled a Greek drama. A portrait of Newton hung on one wall. Scientists sat in silence, awaiting the figures. Einstein had calculated that the sun's gravity would shift starlight by 1.7 seconds of arc. The actual measurements came in between 1.61 and 1.98. On November 10, 1919, the *New York Times* headlined the story:

LIGHTS ALL ASKEW IN THE HEAVENS. . .
Einstein Theory Triumphs
Stars not Where They Seemed
or Were Calculated to Be
But Nobody Need Worry

Overnight Einstein was famous. Within a year his rumpled countenance was a symbol of genius rivaling Edison's "little globe of sunshine" twinkling above a human head. "Since the light deflection result became public," Einstein lamented, "such a cult has been made out of me that I feel like a pagan idol." He remained in awe of light. "For the rest of my life," he wrote, "I will reflect on what light is." Never again, however, would he ride on his beam—or wonder about light—alone.

On into the 1920s, light quanta revived the problem the nineteenth century thought it had solved. Particle or wave? The radical changes suggested by light quanta were put to test after test. Unlike the ether, they passed. Late in 1922, in a lab outside St. Louis, Arthur Compton fired X-rays at a piece of graphite. As expected, the X-rays deflected at all angles. But when Compton measured, he found that each deflected ray had lost a precise quantum of energy. The more profound the angle of deflection, the more energy was lost. Like the "photoelectric effect," Compton scattering suggested light to be particles. Yet in every optics lab, Thomas Young's interference lines still showed light to be waves. Even Einstein was uncomfortable: "There are therefore now two theories of light," he observed, "both indispensable, and—as one must admit today despite twenty years of tremendous effort on the part of theoretical physicists—without any logical connection."

In the absence of answers, theories flooded in. In 1924, a French prince turned physicist, the debonair Louis de Broglie, suggested that light consisted of "pilot waves." Like cars on a roller coaster, particles ride waves that propel them along predictable curves. Particle *and* wave. Using everything from algebra to $E = mc^2$, de Broglie calculated how waves propel quanta. Einstein said the French prince had "lifted a corner of the great veil," but the theory was also mocked as "la *Comédie Française*." Meanwhile, Bohr, more comfortable with uncertainty, insisted that light could be particle *or* wave, depending on the experiment. Then scientists at Bell Labs in New York fired electron beams through crystals. Their results—light diffracted like waves but scattered like particles—made "duality" the catchword. Frustration deepened.

In the decades following Einstein's Wonder Year, the evolving quantum theory had created an exclusive academy. Pondering light's complexities now required more than just interferometers and cloud chambers. The prerequisites included advanced degrees in physics, the ability to fill entire blackboards with equations, and a willingness to worry endlessly about light's latest enigma.

"Professor," a colleague told the physicist Wolfgang Pauli one day, "you look very unhappy."

"How can one look happy when he is thinking about the anomalous Zeeman effect?"

Only a score of physicists could handle the math, live with the uncertainty, outlast the long walks and longer talks. By the mid-1920s, this exclusive academy saw the stakes rising. Light's duality challenged the very definition of science. As outlined by Aristotle and the Arabs, modernized during the Scientific Revolution, and refined by centuries of increasingly compulsive types using increasingly precise instruments, science depended on exactitude. There had to be *one* explanation, *one* set of equations, *one* series of experiments, predictable and repeatable, to explain each natural phenomenon. But now light, that oldest of puzzles, was defying all reason. Particle *and* wave? Particle *or* wave, depending on the experiment? Nature did not work that way.

Jokes countered the despair. Physicists said they had to teach wave theory on Mondays, Wednesdays, and Fridays, and particle theory the rest of the week. Light, another quipped, traveled in "wavicles."

But the new academy's founding fathers—Einstein, Bohr, and others—were not amused. Shortly after refining Bohr's atom, Wolfgang Pauli lamented, "Physics at the moment is again very muddled; in any case, for me it is too complicated and I wish I were a film comedian or something of that sort and had never heard anything about physics." Einstein himself was skeptical about quantum theory. Early on, he had sensed "what a *Schweinerei* . . . what a stinking mess it was." Now, as physics' world-famous éminence grise, he stiffened his resistance.

Light's quantum academy met in a few specific locations. Bohr hosted colleagues at his palatial home in Copenhagen. Others met at the Cavendish Lab in Cambridge. And every few years, they gathered in Brussels for the Solvay Conference. Funded by a wealthy Belgian industrialist, the conferences gathered a score of men and usually the French Nobelist Marie Curie to discuss cutting-edge physics. Einstein called the first Solvay Conference a "witches' Sabbath," yet he showed up to defend his quanta. "These discontinuities, which we find so distasteful in Planck's theory, seem really to exist in nature," he told the gathering. In 1927, five conferences later, with quantum theory earning more converts, Einstein squared off against Bohr. At stake was the future of physics, of science, of certainty itself.

Einstein, standing at a blackboard, drew diagrams of his latest thought experiment: light passing through one slit striking a plate at a precise moment. Surely this classic model, updated to include precise measurement, discredited uncertainty. But Bohr, pausing, puffing his pipe, answered that light, whether particle or wave, would shift the plate upon impact, making its position uncertain. The debate continued, Einstein positing two slits, Bohr countering with his latest theory, "complementarity." Unveiled at a recent conference in Italy, complementarity was part science, part epistemology. All we could know about nature, Bohr argued, depended on the question asked. There might be no absolute truth about light. If it behaved like a particle in some experiments, a wave in others, then such whims had to be accepted. "It is wrong to think that the task of physics is to find out how nature is," Bohr stated. "Physics concerns what we can say about nature."

Day after day, the argument deepened. Einstein brought a new thought experiment to each breakfast. Bohr pondered it all day, then countered by dinnertime, only to face another hypothesis over croissants and coffee the next morning. Throughout the conference, Einstein and Bohr sent photons (the newly adopted name for light quanta) beaming through their imaginations, but maddeningly, impossibly, light defied these world historical geniuses. Sometimes wave, sometimes particle, it forced the most rational minds to fall back on a higher authority. "God does not play dice," the deist Einstein fumed. And the atheist Bohr replied, "It cannot be for us to tell God how He is to run the world."

The Solvay debate resembled the ecclesiastical synods of the first millennium where some bishops saw light as God's reflection, others as divine messenger or metaphor. Einstein and Bohr came to no conclusions and left the conference exhausted and bitter. Bohr's "soothing philosophy—or religion?—is so finely chiseled that it provides a soft pillow for believers . . ." Einstein said. "This religion does damned little for me." Yet come the next Solvay conference in 1930, the two men went at it again. "This epistemology-soaked orgy ought to come to an end," Einstein lamented, yet it would continue until his death. Meanwhile others would learn to accept duality, uncertainty, and light parsed into quanta. "We have to live with quantum theory," Max Planck said in his final years. "And believe me, it will expand."

Yet Einstein could not shake the psychic tension. Quantum theory worked on paper, he insisted, yet still it was incomplete. "But what is light really?" he asked in a textbook he co-wrote in 1938. "Is it a wave or a shower of photons? . . . It seems as though we must use sometimes the one theory and sometimes the other, while at times we may use either. We are faced with a new kind of difficulty. We have two contradictory pictures of reality; separately neither of them fully explains the phenomena of light, but together they do!"

Wielding *c* like a wand, Einstein had toppled Newton, quantized physics, and destroyed the constancy of mass, energy, space, time, and more. And there would be more, so much more soon discovered. How photons can be converted into positive and negative electrons. How the "spin" of photons emitted in pairs remains coordinated, even across space, one photon somehow signaling the other that its spin has been changed. "Spooky action at a distance," Einstein called this. And Max Planck again—"We have to live with quantum theory . . ."

Late in the autumn of his seventy-third year, Einstein wrote his old friend from the patent office, Michele Besso. "All these fifty years of pondering," Einstein confessed, "have not brought me any closer to answering the question—what are light quanta? Nowadays, every Tom, Dick, and Harry thinks he knows it but he is mistaken." Baffling even Einstein, quantum light was not framed by scripture, art, or cathedral windows. Few theorists in the academy gave much thought to how it might ever be used. Yet within five years of Einstein's death in 1955, quantum light stopped being a theory.

Coming out of the lab, a new, dangerous, and dazzling light beamed into human consciousness. At first, no one was even sure what to call it.

June 1925, Heligoland: Quantum physics' blond boy wonder is suffering from a bad bout of hay fever. Sniffling and wheezing, the twenty-four-year-old Werner Heisenberg has left his teaching post in Göttingen for Heligoland, a rocky island thirty miles off Germany's North Sea coast. Alone in a room overlooking the beach, he continues the calculations that have dogged him for months. As he has discussed on long walks through Copenhagen, mathematics can no longer predict quantum behavior. His beloved numbers just won't cooperate. To pinpoint the location of a particle in a cloud chamber, Heisenberg must *see* it. To see it, he must use light. But like light dislodging electrons in Einstein's photoelectric effect, a beam shifts the particle ever so slightly. The more accurately you judge the position of an electron, Heisenberg concludes, the more uncertainty you give its momentum. Hence, no one can be sure of both position and momentum. For electrons, photons, and all other particles, probability must replace certainty, and certainty must forever be elusive.

On into the night, Heisenberg, who began learning Greek at five and calculus at twelve, uses ultracomplex matrix mathematics to calculate how light makes certainty impossible. By three A.M., he has opened the door to his masterpiece: the uncertainty principle. Alone, ecstatic, feeling as if he is "looking at a strangely beautiful interior," he steps outside. He listens to the waves breaking, then walks to a promontory overlooking the sea, climbs, and sits. Gazing over the water, up at the stars, he waits for sunrise.

CHAPTER 14

"Catching Up with Our Dreams": Lasers and Other Everyday Wonders

O Light that none can name, for it is altogether nameless.
O Light with many names, for it is at work in all things . . .
How do you mingle yourself with grass?

—SAINT SYMEON, "HYMNS OF DIVINE LOVE"

Given the sun's lordly reign over humanity, the discovery of how its light is made should have been trumpeted like the dawn. Had the Greeks unveiled it, the knowledge would have been immortal. Had Galileo fulfilled his final dream—being sequestered in exchange for learning "what light is"—all Europe would have marveled. Under Newton's name, such a revelation would have crowned *Opticks*. But none of these early students had the necessary tool—quantum theory. And by the time the quanta explained the sun, the news was hardly noticed.

On August 15, 1938, as the world marched toward another war, a six-page article appeared in the *Physical Review*, the prestigious journal of the American Physical Society. The article's title, "The Formation of Deuterons by Proton Combination," did not mention the sun. Its primary author, Hans Bethe, was unknown to the public. The math was relentless, and the concepts, though starting with basic chemistry, quickly expanded into calculus, quantum mechanics, wave functions, and worse. The sun, Bethe wrote, is an atomic furnace fusing hydrogen into helium, releasing light as a byproduct. This idea had surfaced in the 1920s, yet skeptics doubted the sun was hot enough to sustain such fusion. To this suggestion the British physicist Sir Arthur Eddington, who first proposed the idea, answered, "Tell them to go find a hotter place!" But even if the sun had the necessary heat, others argued, it did not have enough hydrogen. Then a young British woman at the Harvard Astronomical Observatory showed otherwise. Cecilia Payne's Ph.D. thesis on "Stellar Atmospheres" argued that stars contain common elements found on Earth but have far more hydrogen than anything else. Years of gentlemanly "ahems" followed. In the meantime, "quantum tunneling," where, in defiance of classical physics, a particle passes through a seemingly impenetrable force-barrier, provided another step. It remained only for someone to do the math.

Though later known for his work on the Manhattan Project and nuclear arms control, Hans Bethe was only beginning his career when he deciphered the sun. A burly man with a crew cut, bright blue eyes, and a booming laugh, Bethe had fled the Nazis to find a lifelong home at Cornell University. In the spring of 1938 he answered a challenge issued at a conference on stellar astronomy. The previous half century had seen astronomy and physics merge to explain much of celestial light. Light had become a measurement, the light-year denoting the astounding distance, some six trillion miles, that light travels in a year. Astrophysicists had calculated the pulses of variable stars, classified stars from dwarfs to supergiants, and discovered the red shift of light coming from galaxies receding madly from us. But ordinary sunlight? Who among the distinguished physicists at the conference could explain it? "People were really at a loss as to what to do and what reactions to consider," Bethe remembered. Amazed at the "total ignorance," he began working fifteen-hour days, alone with pencil, paper, and slide rule. Solving the eternal riddle required finding the exact ingredients—chemical elements and their particles and subatomic particles—that would fuse at precise temperatures and burn for billions of years. Bethe had to work out the reactions, the quantum tunneling, and the quantities of hydrogen involved. By early summer, after consulting with the physicist Charles Critchfield at George Washington University, Bethe had the recipe for the light of the universe.

Preheat the oven to twenty million degrees Kelvin. Take two nuclei of hydrogen, each consisting of a single proton. Mold with the pressure found only at the core of a star. Let the temperature and pressure overcome the protons' mutual repulsion, forcing them to collide. Let protons fuse into deuterons—heavy hydrogen—releasing a neutrino (an electrically neutral subatomic particle), plus one electron and its quantum-based opposite, a positron. Stand back and cue the brass section. Electron and positron soon collide, destroying each other and emitting a single photon—light. But the photon is just the beginning. To keep the reaction going, add carbon, whose catalytic properties enable the formation of helium. On through reaction after reaction, let protons, positrons, and electrons smash, releasing more photons, fusing into more complex nuclei. Serve by sending photons from the center of the sun to your table.

Hans Bethe did not calculate the photons' wayward path out of the sun's center, but the idea's original proponent, Arthur Eddington, imagined a colorful travelogue:

> The inside of a star is a hurly-burly of atoms, electrons and [radiation] . . . Try to picture the tumult! Disheveled atoms tear along at 50 miles a second with only a few tatters left of their elaborate cloaks of electrons torn from them in the scrimmage. The lost electrons are speeding a hundred times faster to find new resting places. Look out! There is nearly a collision as an electron approaches an atomic nucleus; but putting on speed it sweeps round it in a sharp curve . . . Then comes a worse slip than usual; the electron is fairly caught and attached to the atom, and its career of freedom is at an end. But only for an instant. Barely has the atom arranged the new scalp on its girdle when a quantum of [light] runs into it. With a great explosion the electron is off again for further adventure.

Each second, the sun converts four million tons of mass into pure energy. The photons made in the process can spend up to a million years ricocheting through the furnace, a journey scientists call a "random walk." Once reaching the sun's surface, photons zip to Earth in just over eight minutes. After such a voyage, all of humanity's tributes to sunlight—from the *Upanishads*' "shouts and hurrahs" to Monet's *Impressionism, Sunrise*—seem the least a grateful humanity can offer.

But by 1938, the developed world was plugged in and switched on. Industrialized societies awoke well after dawn and were too busy for sunsets. Advancing around the Earth, darkness made houses glow and turned cities into candelabras. With night no longer a menace, day no longer a sanctuary, light had won its primordial battle with darkness. "Light," the British physicist William Bragg proclaimed, "brings us news of the universe," yet even schoolchildren knew that starlight was old news. Any visible star might have burned out long ago, its ancient light just now reaching us. With starlight fading into the glare of the modern world, Hans Bethe's recipe for sunlight caused little fanfare. Aside from a few news items, the answer to the persistent puzzle of how light is made went straight into the science books. Cynics repeated the old adage "There is no new thing under the sun," yet by the time Bethe finally received a Nobel Prize for his breakthrough, in 1967, there was something utterly new, a light brighter than the sun.

The idea behind the laser came from Einstein, of course. Back in 1916, finished with relativity yet still wondering about light, Einstein pondered Niels Bohr's solar system atom. Bohr had said that quanta of light were emitted spontaneously when excited electrons lost energy and leapt back to their steady inner orbits. Einstein took the premise further. With enough stimulation, he realized, electrons would emit photons that would excite other electrons to emit more photons. And these new photons would excite electrons to emit still more photons exciting more electrons. . . . Along with Bohr's "spontaneous emission," Einstein foresaw "stimulated emission" of light. Pump atoms with enough electromagnetic energy and you could create a beam so powerful that the sun would pale by comparison. Einstein did not live to see the first laser, yet a year before his death in 1955, the laser's stepfather was born.

Because radar helped win World War II, the Pentagon was interested in boosting its microwaves. Shorter, stronger waves would bounce off smaller objects and back to the spinning dish, improving detection of ships and planes. After the war, the Defense Department funded research on stimulated emission with microwaves. Pumping up these longer waves, causing them to pump each other, physicists Charles Townes and Arthur Schawlow made the first maser in 1954. The name, settled upon after Greek and Latin labels seemed contrived, stood for Microwave Amplification by Stimulated Emission of Radiation.

Physicists knew that light waves could also be amplified, but no one expected to do so anytime soon. The goal would be the same achieved with masers—a "population inversion" in which far more electrons were stimulated than stable, pumping each other to higher and higher energy levels. But light waves are ten thousand times as compact as microwaves, and would need ten thousand times the energy as a jumpstart. Even maser pioneer Charles Townes did not expect stimulated light emission for another twenty-five years. Still, the plans were drawn up. Seems the Pentagon was interested not just in radar but in a dream as old as burning mirrors.

Ever since Archimedes allegedly ignited ships by reflecting the sun, light had been sci-fi's favorite "death ray." In 1809, Washington Irving's fantastical *History of New York* imagined invading moon men "armed with concentrated sunbeams." Come the age of comic books and movies, space heroes Buck Rogers and Flash Gordon fired bullets of pure light. And in the fall of 1938, Americans listened on radio as Martians, said to have landed in New Jersey, attacked with rays that were projected "much as the mirror of a lighthouse projects a beam of light." Orson Welles's "War of the Worlds" spawned spin-offs galore, until, by the 1950s, the *zapppp* of "blasters" and "ray guns" was a fixture of cheap sci-fi movies. *This Island Earth. Forbidden Planet. The Day the Earth Stood Still.* And so on. Reality came a step closer to science fiction in 1958, when the maser inventors Townes and Schawlow wrote a seminal paper entitled "Infrared and Optical Masers." The paper fired what the laser historian Jeff Hecht called "the starting gun." Competitors in the race for the first laser included IBM, Westinghouse, Columbia, and MIT. But a small Long Island firm, TRG, had two advantages: a million-dollar Pentagon grant and Gordon Gould.

With his pasty face and nerdy glasses, Gordon Gould seemed to have stepped from one of those fifties sci-fi movies. As a grad student at Columbia, Gould began toying with stimulated emission. Finally, in the fall of 1957, after working for days without stepping outside, Gould took out a legal pad and sketched a long rectangle with its short ends protruding. Above the crude design he wrote, "Conceive a tube terminated by optically flat, partially reflecting terminal mirrors." Gould's tube resembled the neon lighting that had become standard in bars and restaurants, its eerie glow made by excited electrons of neon gas. But Gould saw that if the tube were opaque and lined with mirrors, trapped photons would, at the speed of light, whip up "stimulated emission" whose beam could "heat an object up to 100 million degrees." Gould headed his concept "Some rough calculations on the feasibility of a LASER: Light Amplification by Stimulated Emission of Radiation." It was the first use of the word.

Gould might have earned a patent had he known that a design alone was sufficient to apply, yet thinking he needed a device, he went back to work. Soon hired by TRG, he made design leaps that would lead to numerous lawsuits and four dozen patents, making Gordon Gould both bitter and rich. But Gould's leftist politics, including work for universal disarmament, led to his being denied the security clearance he needed to lead Pentagon-funded research. Others also floundered, pursuing problems that outnumbered solutions. How to oscillate light in a tube yet finally release it in a beam? How to excite a critical mass of electrons, given their disturbing tendency to decay into steady states? How to find the right material for stimulated emission, one that would absorb a wide range of light waves yet emit only one frequency? The material had to be manageable without a roomful of grad students twisting dials, and it had to be stable under intense bombardment of electromagnetic energy. Like Edison searching for the right filament, physicists tried element after element. The most likely candidates were gases—potassium vapor, cesium vapor, mixtures of krypton and mercury, or helium and neon.

Theodore Maiman had a different idea. The son of an engineer, Maiman had conducted his first optical experiment at age three. To show to his mother that the refrigerator light was not turning off, Maiman crawled inside and shut the door. He escaped somehow, having proved his point. Skinny and hyperactive, Maiman grew up tinkering in a basement lab strewn with tubes, batteries, and bulbs. Getting his first engineering job at seventeen, Maiman found physics at Columbia boring. In 1952 he hitched cross-country to Stanford and talked his way into its graduate physics program. After earning his Ph.D., then setting out on a solo trip around the world, Maiman finally returned to his native Los Angeles. Hired by Hughes Laboratories in 1956, he joined what he later called "the technological Olympics" in pursuit of the laser. Maiman had no Pentagon funding. While his rivals led teams of engineers, he worked with a single grad student. But independence gave him the freedom to innovate. Others might try gases, but a crystal, Maiman noted, had a "relatively high gain coefficient," meaning enough atoms could be amped up using a small amount of material. Even a marble-sized chunk of ruby, synthetically grown to eliminate impurities, might work. Maiman was well into his design when a Bell Labs engineer visited Hughes's new facility in Malibu, California. "We hear that you are still working on ruby," he told Maiman. "We have thoroughly checked out ruby as a laser candidate. It's not workable." But Maiman had already reviewed calculations made at Bell. They used the wrong wavelength for the ruby's photon emissions.

In the spring of 1960, colleagues at Hughes found Maiman "just single-minded in his devotion to this thing." Some scoffed: "What would Hughes do with a laser?" Maiman had no answer. The uses would come later—a few thousand of them. First there were more problems to solve. To make his little ruby both reflect and emit light, he coated each end with silver, then made a pinpoint hole in one shiny swatch. The silver would serve as parallel mirrors; the hole would emit the first laser beam. Now all Maiman needed was a light, as blinding as possible, to ignite the revolution. "Optical pumping," some called the process.

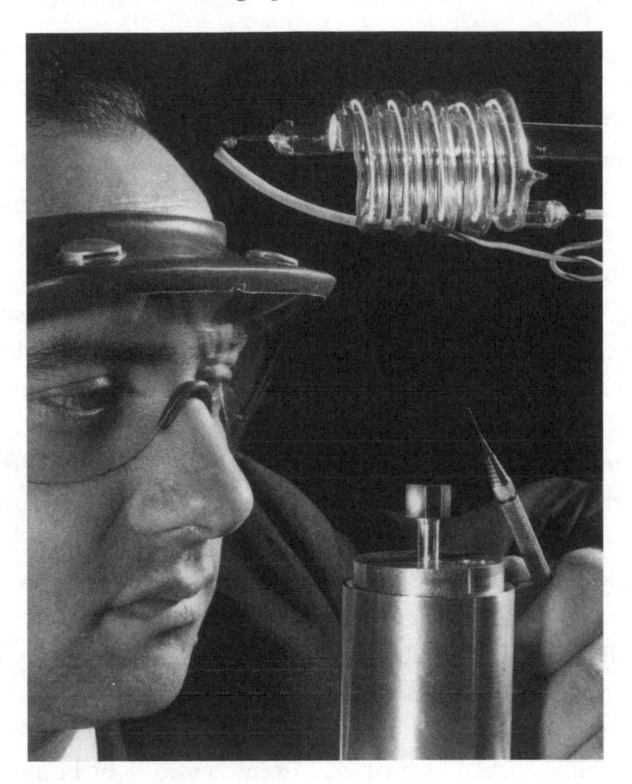

In May 1960, Theodore Maiman won the race to make the first laser. Maiman's device used a chunk of ruby, a mirror-lined tube, and a flash bulb. Courtesy of HRL Laboratories, Malibu, CA

To create the necessary "population inversion," Maiman needed a flash that would excite more than half the ruby's atoms. He considered arc lamps like those that had offended Robert Louis Stevenson—they had become Hollywood's klieg lights. Too unwieldy. He checked out movie projector lamps, one so bright it had to be air cooled. "A nasty lamp to work with," he concluded. Finally he looked into photographers' strobe lights. When his assistant showed him a bulb whose flash temperature topped 7,700° Celsius, Maiman had his "aha!" Leafing through a photography catalogue, he found the model—a shop-ready GE flash bulb. Coiled like one of today's compact fluorescents, the bulb surrounded the chunk of ruby. Hooking the bulb to a power source, adding a capacitor that contained the voltage until it was enough to ignite the flash, Maiman set his small device in a mirror-lined cylinder. At four P.M. on May 16, 1960, light's second Creation began.

The windowless laboratory was stacked with boxy electronics, laced with black cables, lit by ceiling fluorescents. Elsewhere, at enormous East

Coast research labs, engineers were pumping amplified light through three-foot glass tubes hooked to machines the size of suitcases. Theodore Maiman's laser was no bigger than a drinking glass. With his assistant ready, he dialed up the voltage. A green line traced across the oscilloscope's circular screen. When the voltage reached five hundred, Maiman fired the flash bulb. The green line snaked—just slightly. He cranked the dial again. At nine hundred volts, the line jumped. And "when we got past 950 volts on the power supply, everything changed!" Maiman remembered. "The output trace started to shoot up in peak intensity and the initial decay time rapidly decreased." A deep red light filled the room. Maiman's assistant was color-blind, unable to see red on a printed page, but he saw this red and jumped. Maiman, "numb and emotionally drained," sat in silence. Colleagues came to see the newborn light. The first laser was not continuous—that leap would soon be made at Bell Labs—but its pulses had broken the barrier. Einstein had again been proved prescient. Hughes officials wanted an immediate press release, but Maiman insisted on further testing.

On July 7, 1960, the public heard the news. The discovery of how the sun shines had barely been noticed, but the "optical maser" earned a press conference at Delmonico's Restaurant in Manhattan. Human ingenuity, Hughes announced, had created an "atomic radio light brighter than the center of the sun." Reporters puzzled over this "coherent light." Unlike scattered sunlight, they learned, laser light is perfectly aligned, its monochrome waves all in sync, packing its power into filaments as thin as twenty-seven millionths of an inch. Journalists were skeptical or confused, but engineers at Bell and other labs understood—they had lost the race. Reporters listened as Maiman proposed future uses: long distance communications, cutting, welding, even surgery. Then one asked whether this "coherent light" might become a "death ray." Not for at least twenty years, Maiman replied, disgusted by the question. The following day the *Los Angeles Herald* ran a two-inch headline: LA MAN DISCOVERS SCIENCE FICTION DEATH RAY.

Magazines echoed the ancient fear. "Bullets of Light" (*New Republic*). "Ruby Ray Guns" (*America*). "Light Ray: Fantastic Weapon of the Future?" (*U.S. News and World Report*). The fantasies were trumped only by the truth as each year brought another marvel.

- December, 1960: Using the first continuous laser, Bell engineers send a message on an optical beam. In tribute to the telephone's creation story, the voice says, "Come here, Watson, I need you."

- 1961: Doctors at Columbia Presbyterian Medical Center in Manhattan use a laser to burn out a retinal tumor.
- May 9, 1962: Using a ruby laser, MIT's Project Luna See bounces a beam of light off the moon.
- 1963: In a spin-off of the Michelson-Morley experiment, two MIT scientists use lasers to try to detect luminiferous ether. It still doesn't exist.
- 1964: At the World's Fair in New York, the AT&T Pavilion shows how a single laser beam will someday carry 10 million phone calls.
- 1965: Gemini 7 astronaut James Lovell, holding a six-pound laser, beams his voice one hundred miles down to a NASA tracking station in Hawaii.

And always, the echo of Archimedes. General Curtis LeMay, ignoring the doubts of Maiman and other physicists, announced that the Air Force was pursuing "beam-directed energy weapons" to shoot down Soviet missiles. "Our national security may depend on weapons far different than any we know today," the cigar-chomping LeMay said. "Perhaps they will be weapons that strike with the speed of light . . . With such a capability an enemy would have the potential to dominate the world."

In an age of techno-wonders—transistors and Telstar and space flight—lasers also cut a swath through pop culture. In the 1964 movie *Goldfinger*, James Bond lies strapped to a table, a sinister ray gun aimed between his legs. "You are looking at an industrial laser," the villainous Goldfinger tells Bond. "It emits an extraordinary light, not to be found in nature. It can project a spot on the moon. Or at closer range, cut through solid metal. I will show you." With a *zappp*, the ruby beam slices a smoldering slit that inches toward 007's privates. (Bond escapes.) Within a year, TV's *My Favorite Martian* is building a laser in his garage. Within two, Captain James T. Kirk and the crew of the starship *Enterprise* are setting their "phasers to stun."

In September 1967, CBS's weekly program *The 21st Century* aired "Laser: A Light Fantastic." "Since the universe began," Walter Cronkite announced over a flickering candle, "light has remained unchanged. Now man has created a new kind of light with powers and properties unlike anything that existed before—laser light." Against a backdrop of red and green shadows, Cronkite imagined the possibilities—picture phones, microsurgery, personal computers, the world's communications transmitted in a single beam. "The laser today stands on a technological frontier,"

Cronkite concluded, "a frontier we're only beginning to explore. Where the laser's fantastic light will shine is a challenge we face as we move into the twenty-first century." But the wait was not that long. By 1970, lasers were welding integrated circuits, guiding bombs dropped on North Vietnam, mending detached retinas, and measuring the distance to the moon within an inch. Maser pioneer Charles Townes: "We are catching up with our dreams."

Throughout the 1970s, as lasers proliferated, residents of Pasadena, California, puzzled over an unusual Ford van. Spotted throughout the city, the brown Econoline was decorated with circles, squiggles, and angular lines. Every now and then some physics grad student would stop the driver to ask, "Why do you have Feynman diagrams all over your van?" And the driver, beaming, would reply, "Because I'm Richard Feynman."

Richard P. Feynman loved light more for its quirks than for its radiance. While others raged at the improbabilities of quantum theory, Feynman relished them. A curious mix of New York street sass and Ivy League Ph.D., Feynman peppered his physics lectures with words like "screwy," "dopey," and "absurd." "There was a moment when I knew how Nature worked," he once said. "It had elegance and beauty. The goddamn thing was gleaming." From early in his career, Feynman's playfulness added to legends about his intellect. Back in 1943, fresh out of Princeton and plunged into the Manhattan Project, Feynman spent his spare time at Los Alamos stalking top secret labs learning to crack safes. Just for fun. Though still in his mid-twenties, he also gave exuberant talks that drew the whole pantheon present at Los Alamos: Niels Bohr, Hans Bethe, J. Robert Oppenheimer and others. Oppenheimer recognized Feynman as "the most brilliant young physicist here." What few knew was that on weekends, Feynman was hitching to Albuquerque to visit his dying wife in a sanitarium. When Arlene succumbed to tuberculosis shortly before the bomb was finished, the distraught Feynman doubled down on his resolve to have boyish fun all his life. For the next four decades, dancing and drumming in nightclubs, delighting students with his witty lectures, he and the quantum light he fine-tuned seemed determined to defy expectations. "He himself," one colleague wrote, "behaved much like the electron."

Back from Los Alamos after the war, Feynman joined the boisterous Hans Bethe at Cornell. There both professors puzzled over enigmas that had persisted since the Einstein-Bohr debates. At microscopic levels, when photons interacted with matter, when light "mingled with the grass," the math broke down. Bethe recalled the confusion:

> The idea is that you have the quantum theory of electrons, which shows you how an atom is built, and shows you how to calculate the energy levels of an atom. Then you have the quantum theory of radiation . . . The problem was to put these two things together, and you had to do it paying attention to special relativity. The trouble was that this quantum electrodynamics worked very well if you calculated everything only in first approximation. If you tried to do it more accurately, it gave the result: infinity. Which was obviously wrong . . . So a method had to be found to get rid of the infinities.

Bethe tried and failed. Others, including Julian Schwinger at Harvard and Sin-Itiro Tomonaga of the Tokyo University of Education, made mind-numbing calculations that seemed to work. Then, in 1949, Richard Feynman picked up a pen, drummed his fingers as he always did when calculating, and drew the goddamn thing.

Niels Bohr's solar-system-like atom remained a fixture in textbooks, but physicists had long since refined it. Electrons, it seems, orbit nuclei in waves, emitting light in ways no one could picture—until the Feynman diagrams. Feynman drew photons as squiggles, electrons as arrows. He numbered their collisions in sequence: #1-electron hits photon . . . #2-photon spins off . . . With a vertical y-axis representing the passing of time, one could place paper over the diagram, slide it upward and as if in animation, watch light and matter mingle. The Feynman diagrams showed light and matter collide according to carefully calculated probabilities. But the diagrams looked more like spaghetti than any picture of the cosmos. Nature can't be this weird, some said, nor so simply explained, but diagram after diagram unfolded, not describing but *showing* the emission and absorption of photons, the scattering of light, the dispersion of light.

Feynman dismissed his diagrams as some "half-assedly thought-out pictorial semi-vision thing." But after he shared the Nobel Prize for Physics in 1965, he painted the lines and squiggles on his van and drove around his adopted home, Caltech in Pasadena, with a license plate that read QANTUM (California vanity plates only allowed six letters). Though the Feynman diagrams appeared childlike, one physicist called them "the sun breaking through the clouds, complete with rainbow and pot of gold." Using the

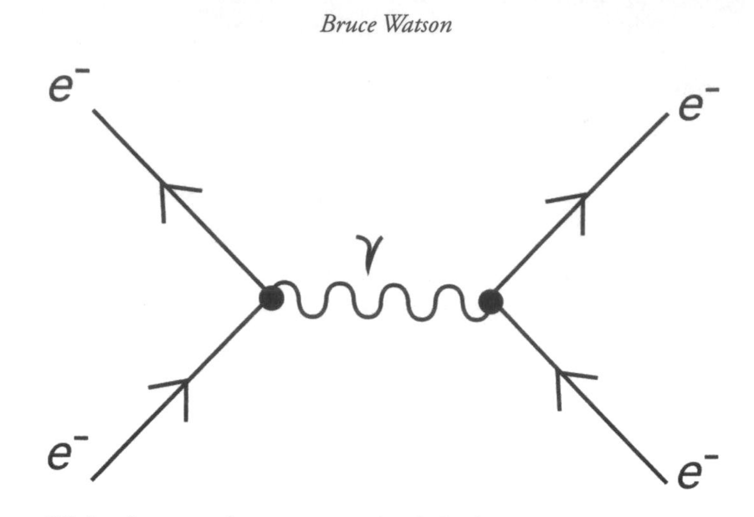

With a few squiggles, arrows, and a dash of genius, physicist Richard Feynman diagrammed how light interacts with matter. His Feynman diagrams remain a key part of quantum electrodynamics.

diagrams, a physicist did not have to be a Maxwell or a Feynman to *see* how light interacted with matter. Equally golden was Feynman's QED, "quantum electrodynamics, the strange theory of light and matter." One needed to be a "calculating machine" to fully grasp QED, especially when Feynman added "probability amplitudes," little arrows whose length and direction modeled the likelihood of a photon's path.

But the math behind QED was uncannily accurate. Feynman often boasted that if the distance from New York to L.A. were measured with the precision of QED, the measurement "would be exact to the thickness of a human hair." And whatever mysteries of light Feynman could not explain he celebrated. QED, he told audiences, was "the jewel of physics," able to describe the most intricate interactions of light and matter—how oil slicks make rainbows, how light reflects off glass yet also beams right through. Feynman admitted however, that QED "describes Nature as absurd from the point of view of common sense . . . So I hope you can accept Nature as She is—absurd."

QED took a generation to sink in, with many more physicists refining the math and expanding the list of light's quirky behaviors. But once lasers filled every optics lab, Feynman's blend of relativity and quantum theory explained precisely how light behaved—except when it couldn't. Consider how the quantum messed with Thomas Young's two-slit experiment. Beam light through two slits and it makes stripes, the familiar interference patterns. But beam single photons through the slits and they still interfere, just as if each were its own full wave. How does a photon "make up its mind" which slit to choose? Feynman asked. Why does a single photon interfere with another? Still more "screwy" things happen when you put detectors at each slit to chart the path of each photon. Then the interference patterns disappear altogether, leaving you with lumps of photons on your screen. Does light resent being "watched"? With a sly grin, Feynman offered a ready answer: "Do not keep asking yourself, if you can possibly avoid it, 'but how can it be like that?' Nobody knows how it can be like that."

By the mid-1970s, with QED as the bedrock of "quantum optics," a new field of light was born—photonics—the branch of physics studying photons' uses in technology, especially communications. Photonics engineers created the latest lasers, ranging from free-electron lasers, which amplify light at all frequencies, to laser diodes so small they fit in a pocket, and so cheap they can be mass-produced to read compact discs, DVDs, and bar codes at your local supermarket. Quantum optics helped reduce random "noise" as photons careened through longer and longer "light pipes," a.k.a. fiber-optic cables. Though far more intricate than Newton's *Opticks* or Fresnel's integrals, quantum optics became the holy grail of physics graduate programs. And the fun of playing with light, coupled with rising salaries paid by firms tapping the multi-billion dollar laser market, drew men and women from around the world. What they found was a vocabulary as thick as any created by light's earliest disciples.

The typical quantum optics textbook is subtitled "an introduction," but after a whirlwind review of "classical optics" from Euclid to Maxwell, the equations begin in chapter 2. Then this brave new light unfolds with integrals and matrices far outnumbering actual sentences. Deeper and deeper the quantum probes, revealing the most arcane aspects of light—photon antibunching, high-order photon coherence, zero-point energy, chaotic light, squeezed light, elastic and inelastic light, quantum phase gates . . . Quantum optics labs demand the most advanced equipment: hypersensitive photomultipliers, fusion splicers, photodiode amplifiers, and lasers amping up light from different gases.

Quantum optics also entangled the English language. Thomas Edison did not call his globe of sunshine an IB-CBF (Incandescent Bulb with Carbon-Based filament). The public would have cringed and relit the gas lamps. Yet the laser ushered in a new era in naming electronic devices. The simple CD (compact disc) begat the CD-ROM (compact disc—read-only memory). DVD (digital video disc) did not grate the ear and then came ladar (laser detection and ranging), CCD (charge-coupled devices used in telescopes), and the corrective eye surgery called Lasik (laser-assisted in situ keratomileusis). And these were just the acronyms used for commercial products. Students of quantum optics must learn BPP (beam parameter product), FROG (frequency-resolved optical grating), DBR (distributed Bragg reflector), REMPI (resonance-enhanced multiphoton ionization), and GRENOUILLE (grating-eliminated no-nonsense observation of ultrafast-incident laser-light E fields). Webster wept.

Amped up by quantum optics, light's late-century miracles began with "laser cooling," using light to bombard atoms, slowing them until temperatures approach absolute zero, the coldest anything can get. Lasers also worked as tweezers, their angled beams "holding" individual atoms. In industry, lasers cleaned corroded materials, welded better than any flame, and etched microchips with nanometric precision, enabling a single chip to contain billions of transistors. In medicine, lasers performed microsurgery and blasted cancerous cells. Yet they also fine-tune human skin, removing tattoos or unwanted hair. Laser scanners created 3-D images of buildings and bridges that rotated on computer screens. And still the Archimedean dream remained. Since 1983 the Pentagon has pursued President Ronald Reagan's quixotic vision of space-based laser weapons to *zap!* incoming missiles. Three decades and $150 billion later, some sort of breakthrough is (always) expected soon.

By the dawn of the twenty-first century light had come of age. The curiosity of an ancient Greek eccentric in a purple robe had blossomed into a vast field of study replete with enigmas and endless innovation. Light remained pure energy, as Maxwell had discovered, yet that energy was no longer just electromagnetic. Light energized billion-dollar industries and pumped out Ph.D.s and postdocs. Light burst from every hand-held gadget and danced through laser-lit arenas. Seen from Earth, light bleached the night sky; seen from space, its yellow Rorschach splotches defined countries rich and poor, rural and urban. There seemed no limit to what light might do, what it could promise, where it might lead us.

In the beginning, light was God or else His emissary. Each dawn inspired myth and worship, ritual and song. We lifted our eyes to the light and forgot our fears in its presence. With time, some learned to catch light with mirrors, to diagram its rays, to channel it through stone and gather it with lenses. Others wove it into metaphor or captured it on canvas. None could imagine its speed and all were astonished by its beauty. But by the end of the twentieth century, the quantum century, all that eulogy, all that inquiry seemed so innocent, so long ago. Though sun and moon still made their rounds, though stars still sparkled, light had come under human control. Physicists spoke of using light to create nuclear fusion like the sun's own. Light, some said, would power the next generation of computers, their circuitry run by photons instead of electrons. Some even spoke of using light's entanglement—Einstein's "spooky action at a distance"—to teleport matter as in "Beam me up, Scotty." But far from labs and classrooms, the eternal question still burned.

What is light? After some four thousand years of enchantment, the answer, as in quantum wave-particle experiments, varies with expectations. To artists, light is the maker of shadow and the acid test of talent. To the devout, light remains divine, while to the dying, it can seem the portal to paradise. To physicists, light is that clever rope trick of entwined waves, one electric, the other magnetic, together shining like the stars. And to photonics engineers and all who use their everyday miracles, light is the most powerful and precise of all tools, healing, burning, measuring, etching, reading, singing . . . Now in its fourth millennium of inspiring curiosity and awe, light remains what it was in the beginning—the magician of the cosmos.

Epilogue

Ye are all the children of light, and the children of the day: we are not of the night, nor of darkness.

—I THESSALONIANS 5:5

You wake with the sun, squinting into the spreading radiance. Your clock's LED readout tells you the time—7:01. Dressed and in a hurry, you use a derivative of the maser—the microwave oven—to heat water for coffee. Your CD player, its laser diode decoding tiny etchings in the compact disc, plays your morning soundtrack. Wolfing down your English muffin, you read the newspaper through the polarized pixels of your laptop screen. When the microwave's LED shows 8:00, you check your smartphone, its light spelling out the day's appointments. Driving to work, you listen to another CD while your car uses adaptive cruise control to adjust its speed according to laser readouts of traffic.

Once at work, you boot up your computer, making light your workhorse. On your desktop, the World Wide Web, beamed through fiber-optic filaments, brings news of the universe. Another laser diode traces every movement of your computer's mouse. Every document goes to your laser printer. Later, speaking to your team, you hold a three-volt laser in your hand, its red pinpoint flitting across a spreadsheet of pure light. On your way home, when you stop to buy groceries, each item's price is read by laser from a bar code. You pay with a credit card etched with a rainbow hologram. Back home, your laptop and CD and DVD players all trip the light fantastic.

And when the digital clock tells you it's late, you turn out the compact fluorescent by your bedside. Above you, the great wheel of stars that entranced your ancestors spins on, faded now by the floodlit night.

Because it is eternal, there is no end to light. Photons, unlike other subatomic particles, have no mass and therefore do not decay. Whether begat by God or an indifferent cosmos, the first photons of creation are still out there—somewhere. (Cooled over billions of years, the first photons are part of the cosmic background radiation left over from the Big Bang.) Reverence for light seems equally everlasting. Though science has mastered light, throngs of worshippers still gather for solstice sunrises—at Stonehenge each summer, at Ireland's Newgrange each winter. In classrooms where children play with lenses and mirrors, in optics labs where lasers thread through thin air, light retains its fascination. This duality—not just particle and wave but mystery and miracle worker—leaves light in the twenty-first century with four incarnations. The Light Enchanting. The Light Intrusive. The Light Astonishing. The Light Eternal.

THE LIGHT ENCHANTING

For a few nights each autumn, the skies over Myanmar, in Southeast Asia, catch fire. The fire begins in open fields where rings of flame blaze beneath huge balloons. As tens of thousands cheer, hot air swells the glowing globes decorated with flags, faces, and scrollwork. Inflating to their full size, fifty feet in diameter, the balloons gently rise into the darkness. With torches and fireworks ringing the night, the scene recalls the Buddhist creation story in which first beings, "self-luminous, moving through the air, glorious," created a "World of Radiance." But the Buddha did not foresee the sparkle of light on the ground, hundreds of smartphones held aloft, each screen framing a balloon as it soars.

Like the great orgies of light that heralded the early age of incandescence, festivals of light are flourishing around the world. The oldest, including Myanmar's Tazaundaing Festival, have religious roots. Throughout Southeast Asia, for five days in October, the night shimmers with candles, flares, and fireworks. Rivaling the splendor of sunrise over the Ganges, Diwali, the Hindu festival of light, makes ancient temples glow and lights flames by the millions. Light means more than dazzle in Diwali, as the *Times of India* suggested: "Regardless of the mythological explanation one prefers, what the festival of lights really stands for today is a reaffirmation of hope." A similar renewal occurs as winter looms and Jews celebrate Hanukkah, lighting menorahs to recall the miracle when a single evening's lamp oil lasted eight nights. But the newest festivals go far beyond candles and lamps.

Picture Berlin's Brandenburg Gate lit with purple columns one night, luminescent roses the next. Imagine parks with trees glowing in citrus colors. Visualize the bulbous domes of a cathedral colored in pastel flowers. Now stroll along streets lined with flashing sculptures, radiant tents, and rainbow pillars. Since 2004, the annual Berlin Festival of Lights has drawn up to a million visitors to the revived German capital for ten days in October. Reviewing a recent festival, the *London Times* described "a mesmerizing city of light . . . and rather trippy sea of colour."

Not to be left behind, Amsterdam now hosts its own annual festival of light. For two weeks in December, tourists and natives board canal boats, shivering as they glide past luminescent sculptures, floodlit sails, and, in the year I visited Rembrandt's house, a glowing white bed floating in the city's red light district. More dazzling yet more infrequent is Light Festival Ghent. Every three years, most recently in 2015, the old Belgian city becomes a fairy tale of ornate facades lit by lasers in splashy colors. Ghent's Cathedral of Light turns a city block into a jeweled nave that would have made Abbot Suger fall to his knees. Europe's oldest light fest, Lyon's Fête de Lumières, has evolved from holy candles to laser-lit Fauvist colors. Hong Kong's Symphony of Lights turns the city skyline into a disco with beams pulsing to the beat of truly obnoxious music. America is only beginning to catch up. Each Thanksgiving, Chicago illuminates its Miracle Mile shopping district with a million lights. Providence, Rhode Island, hosts a more primal light fest. Every other Saturday night in summer, crowds line the Providence River to watch WaterFire, a floating festival of small barges bearing bonfires in iron braziers. And of course, every night brings a light festival to Times Square and the Las Vegas Strip.

But the Light Enchanting is not solely a celebration. The twenty-first century may be laser-lit but the light of the future is the LED (light-emitting diode). Invented in the 1960s, LEDs make light from Niels Bohr's atom, their excited photons jumping the gap between the bulb's electrodes. The first LEDs were red, then green. The search for an intense-wave blue LED, which when combined with the other primary colors would make white, dragged on for decades. Finally, in 1994, two physicists in Japan and a third in the United States drew blue light from indium gallium nitride. Since then, light-emitting diodes have become the artificial light the world has long awaited. Because they waste little heat, LEDs use one twentieth the energy of incandescent bulbs. In a world where almost 20 percent of all electricity is used to produce light, LEDs carry their own brand of hope. In African villages, solar-powered LEDs are already replacing smoky kerosene lamps. In 2014, the Nobel Prize for Physics was awarded to the three inventors of the blue LED, which, the Nobel committee said, "holds great promise for increasing the quality of life for over 1.5 billion people around the world who lack access to electricity grids." The United Nations paid tribute by declaring 2015 the International Year of Light (IYL2015) and its emerging technologies.

By 2030, the U.S. Department of Energy expects LEDs to account for three quarters of all lighting sales. But why wait? You can already buy LED chandeliers, LED candles, lanterns, flashlights, key chains, stick-on strips, a seven-color LED shower head, LED floodlights, floor lamps, headlamps, reading lights, a color-changing mood cube . . . So ubiquitous have LEDs become that you might not be surprised to see a bear roaming the woods carrying an LED flashlight. And the ten-watt bulb in my kitchen is an outright marvel. A circle of LEDs inside a screw-in globe, the bulb burns as brightly as a compact fluorescent bulb but uses just a quarter of the energy and contains no toxic mercury. Though it cost $6.99, the manufacturer says it will last twenty-two years—perhaps the last kitchen bulb I'll ever buy.

Cheaper, brighter, ubiquitous, LEDs are turning a world that once shuddered at sunset into a world that conquers darkness with ease. Still, you don't have to be a Romantic poet to wonder. Is light, to paraphrase Wordsworth, "too much with us now"?

THE LIGHT INTRUSIVE

Ninety minutes before sunrise on January 16, 1994, the earth beneath Northridge, California, began to rumble. Seconds later, Southern California awoke to its perpetual nightmare. Sleepers were thrown from their beds. Picture frames and shelves crashed to the floor. Walls cracked, sidewalks buckled, streetlights winked and went out. Grabbing robes or jackets, terrified parents clutched their children and rushed into the suburban streets. There families huddled together in the pitch black, crying, shaking. Then some looked up at the sky.

As the earth shuddered and stilled, the stars sparkled in silence. "When we got the kids and ran outside," one woman remembered, "we found all our neighbors standing in the street, looking up at the sky and saying, 'Wow!'" By mid-morning, astronomers at Griffith Observatory in Los Angeles were fielding call after call. What was up with "the strange sky"? Had the earthquake changed it somehow? "We finally realized what we were dealing with," the astronomer Ed Krupp told the *Los Angeles Times.* "The quake had knocked out most of the power and people ran outside and they saw the stars. The stars were in fact so unfamiliar and they called us wondering what happened."

Any camper who has gawked at a star-spangled sky only to come home to a washed-out dome knows the problem. Where darkness once ruled there is simply too much light. Light pollution, once just an urban nuisance, now pales the skies in all but the most remote areas. In 2001, when cartographers created "The World Atlas of the Artificial Night Sky Brightness," they found the gauze of night on "a global scale." Even at midnight, two thirds of the world's population cannot see more than a dozen stars. "Mankind," the report concluded, "is proceeding to envelop itself in a luminous fog."

Here in our time, the stars that Shakespeare called "night's candles" are fading before our eyes. In cities, suburbs, and small towns alike, shopping centers glow. Gas stations are so bright you can read while you fill your tank. Glaring neon and fluorescent signs seem designed to rival the Vegas Strip. And in suburban neighborhoods, floodlit houses glow brighter than any ship at sea. Overwhelmed by the Light Intrusive, the darksome monster slinks away, and humanity's visceral connection with the universe is all but gone. Astronomers and stargazers flock to the few remaining areas of darkness, the average urban dweller stays inside, and night, more beige than black, never fully falls.

But those enamored of night are fighting back. Since 1988, the International Dark Sky Association (IDSA) has promoted remedies for light pollution. Sixteen states and several cities, including Tucson, Phoenix, and almost every town in Colorado, have adopted "dark sky" regulations. These limit the lumens of billboards and restrict "glare bombs"—street lights that spew radiance in all directions. Britain's Campaign for Dark Skies and Italy's CieloBuio ("dark sky") are waging a similar battle. And where astronomy is a widespread hobby, locals are dousing their own lights. In southern Iran, the city of Sa'adat Shahr, nicknamed "Astronomy Town," periodically dims streetlights to hold "star parties." In 2001, Flagstaff, Arizona, became the first International Dark Sky Community. The IDSA singled out Flagstaff and its Dark Skies Coalition for promoting strict lighting regulations and dark-sky nights. Dozens of other communities have since joined the IDSA list and more will follow, thanks to an International Dark Sky Week held each April.

Yet the battle for the night is only beginning. And to the dismay of dark-sky lovers, the light of the future, the LED, is not helping their cause. As LEDs replace ordinary streetlights, they will save electricity but increase glare. Though the average eye may not notice, stargazers do. The light from old-fashioned yellow streetlights can be filtered out by telescopes, yet the higher-energy blue wavelengths emitted by each white LED are harder to screen. Blue light also scatters more widely than other frequencies, which explains the sky's most common color. Blue, said the University of Hawaii astronomer Richard Wainscoat, is "the nightmare spectrum." Encouraged by cost efficiency, cities are beginning to install LED streetlights, leaving astronomers very worried. "The blue light really has to be suppressed," Wainscoat told *Astronomy*, "otherwise our view of the night in the future is going to be suppressed."

A more timeless threat to the night comes from our visceral love of light. We simply feel better, safer, with more light. Researchers at the Sandia National Laboratories in New Mexico found that when lighting costs decline, the savings are too often absorbed by people using more light, "essentially a 100 percent rebound in energy use." Drawn to its comfort and beauty, we are still in love with light, still troubled by darkness.

We will never return to unimpeded night, when dusk covered the world with "horrid darkness." We would not want to. But as we master light, will we watch silently as the night sky vanishes? If the stars become tangential to human consciousness, what light will inspire awe?

THE LIGHT ASTONISHING

Split by a spar and twisted by a magnet. Amped up brighter than the sun. Ricocheted through crystals and bounced off the moon. There seemed nothing we could not do with light except stop it. And then, in 2001, a team of Harvard physicists fired a laser into a supercooled cloud of Bose-Einstein condensate (BEC). Predicted by Einstein and Satyendra Nath Bose in 1925 and only created seven decades later, BECs are incredibly dense, incredibly opaque subatomic particles. To slow and perhaps stop light, a Danish physicist, Lene Vestergaard Hau, and her team used one laser to open a Bose-Einstein cloud, then fired another beam into it and slammed its web shut.

Understand me now. The light neither bounced nor bent. It entered the tiny cloud and *stopped.* The miracle called "slow light" promises to extend the distance light can travel in fiber-optic cables. But slow light also suggests that, as the University of Arizona physicist Thomas Miltser told me, "Probably anything you can imagine doing with light can happen."

Imagine using light to cure depression. For decades, seasonal affective disorder (SAD) has affirmed the melancholy that Emily Dickinson knew all too well:

> There's a certain Slant of light,
> Winter Afternoons—
> That oppresses, like the Heft
> Of Cathedral Tunes—

With winter's pale light triggering depression, a variety of "light boxes" are sold as treatments for SAD. Studies have shown that for the chronically depressed, simply sitting beside a bright light for fifteen minutes each morning can help fight the blues. Though once suspect as pseudo-science, light therapy now treats depression, jet lag, sleep disorders, and dementia. The Mayo Clinic has endorsed light therapy as "a proven seasonal affective disorder treatment."

But light boxes seem minor miracles compared to optogenetics. In 2011, researchers at Stanford University used light to calm anxious mice. First Dr. Karl Deisseroth and his team treated each mouse's amygdala—the brain's seat of depression—with opsins, proteins in the retina that absorb photons and trigger the optic nerve. Then the researchers fired blue beams to switch off the primed neurons. The anxious mice calmed, even when placed in open fields where predators might lurk. "They just hunkered down," Deisseroth said. Beyond making mice relax, optogenetics is changing neuroscience. If pulses of light can excite the amygdala, what other functions normally stimulated by intrusive electric probes might be triggered by light? Optogenetics is still in incubation, but the National Institute of Health considers the field "the most revolutionary thing that has happened in neuroscience in the last couple of decades."

Now imagine a laser beam that lasts just a femtosecond—one quadrillionth of a second. Imagine its stimulated emissions creating pressures three times as dense as the sun's center. Such a laser already exists at the National Ignition Facility in Berkeley's Lawrence-Livermore Lab. In 2014, NIF scientists achieved "ignition," the creation of energy through fusion, as in the sun.

The world's most powerful laser heated and compressed hydrogen atoms into helium, releasing more energy than was added to the mix. Though still in its infancy—and requiring huge amounts of energy to power the laser itself—this first ignition fanned the biggest possible dream about light. "If we can harness this fusion reaction," said NIF staff scientist Tammy Ma, "this gives us limitless and sustainable energy for humankind."

Imagine invisibility. If light could somehow be diverted around an object, that object would become invisible. Though it suggests Harry Potter's magical cloak, invisibility is no longer left to witchcraft and wizardry. In 2006, Duke University scientists embedded a small ceramic chip with microscopic circuits whose electromagnetic field bent oncoming microwaves. As with the evolution from maser to laser, the chase was on to make such "metamaterials" divert light waves. Again, the future arrived quickly. By 2008, scientists at the University of California, Berkeley had engineered a composite material with a fishnet weave whose holes were more compact than a red light wave. The material bent some light around it and let other wavelengths pass through. The field of metamaterials is now one of the hottest in optics. Though Harry Potter's invisibility is decades away, metamaterials allow dreamers to . . .

Imagine a computer whose circuits are fired by light. In 2010, photonics engineers at IBM created a semiconductor chip that converts photons to electrons, then uses the charged particles to process data. Four years later, two German scientists activated a transistor not with a charged electron, as usual, but with a single photon. The emerging field of silicon nanophotonics is not expected to replace desktops and laptops, but the advantages of computing with light—cooler, faster, more efficient than electricity—parallel the advances of fiber optics over copper phone wires. Some expect supercomputers in industry and medicine to begin converting to photons by 2020. "I thought I understood everything," one physicist said of the first photon transistor, "and this experiment has completely frazzled my mind." Welcome to the club.

Frazzled by imagining all that light might do, I decided to visit the College of Optical Sciences at the University of Arizona in Tucson. Though the college still gets calls from locals wondering whether someone there can fix broken eyeglasses, it is one of America's top optical research institutes. Busy inventing the future, the school lent me two grad students, Dale Karas and R. Dawson Baker, as guides. The red-bearded Karas and the Texas-gentlemanly Baker, having chosen to study light for both fun and profit, are just two of the college's three hundred-plus students, men and women from dozens of different countries.

Here inside glittering glass buildings overlooking cactus and desert skies, the College of Optical Sciences is advancing the study that dates from ancient Greece and China. While professors scurry to meetings and students huddle around laptops, Karas and Baker steer me past lab tables crammed with the offspring of Albert Michelson's interferometer and Theodore Maiman's laser. Imagine Chevalier's optical shop near the Pont Neuf, its shelves stocked not with glass baubles but with optical coherent microscopes, photomultipliers, fiber-optic sensors, liquid crystal light valves . . .

A modern optics lab costs millions, which explains why the college's corporate sponsorship reads like NASDAQ listings: 3M, Edmund Optics, Optimax, HP, Lockheed-Martin, Raytheon, Hitachi, Canon . . . Talking to professors, however, I find more idealism than commercialism, and no shortage of imagination. Someday, they assure me, we will use light to do the following:

- Take 3D photos of blood samples to screen for disease—with a smartphone
- Not merely sculpt the eye with Lasik but implant self-focusing lenses that will make all eyeglasses obsolete
- Create holographic images of the microscopic world
- Screen for cancers with optical coherence tomography (OCT), a light-based version of ultrasound
- Produce electricity from enormous fields of solar panels, providing up to one sixth of the world's electricity by 2040
- Scan, probe, stimulate, or image almost anything within the human body and within reach of a laser and a lens

The pride of the College of Optical Sciences is its mirror lab. Here, tucked beneath the Arizona Wildcats' football stadium, technicians are crafting mirrors that will revolutionize astronomy.

Though the largest telescope mirrors used to be five, ten, or fifteen feet in diameter, the Steward Observatory Mirror Lab is casting mirrors nearly twenty-eight feet in diameter. Like an enormous magnifying glass, one shiny white disc lies in a massive steel frame on the floor below me. A mirror this large is made by melting chunks of the purest glass into a honeycombed web, then cooling the molten mass one degree a day for six months.

The fine art, however, comes in the polishing, as computerized sanders spend years removing all impurities. If the finished mirror were the size of the U.S., it would be flatter than the Texas panhandle. Any one of these

The largest telescope mirrors ever made, nearly twenty-eight feet in diameter, will enable the Giant Magellan Telescope in Chile to search for the first light of the universe. The mirrors are made in the Steward Observatory Mirror Laboratory at the University of Arizona in Tucson.
Photo by author

mirrors would gather light Galileo never imagined, but astronomers building the Giant Magellan Telescope in Chile will use seven of them. Arranged in a circle of circles and aimed at the stars, the Magellan and other telescopes of the future will see "First Light": the stellar glow emitted from the inaugural generation of galaxies.

THE LIGHT ETERNAL

First Light did not begin with the Big Bang. (Neither did it shine from God's eyes, nor blare as in Haydn's *Creation.*) According to cosmological theories still being refined, the first few seconds of the universe created a stew of elemental materials—electrons and quarks, neutrons and protons—within a seething matrix of ultra-high-energy photons that scattered off free-floating electrons. The result was a plasma of overwhelming brilliance, an all-enveloping, opaque curtain of light. Finally, the plasma cooled enough to allow electrons and protons to form the first atoms, freeing photons to roam unimpeded in a newly transparent universe. But only after another 400 million years—an epoch cosmologists call the Dark Ages—did stars begin to form, streaming light into the cosmos. Let there be questions. Confusion. Wonder.

In January 2011, the Hubble Space Telescope detected a galaxy 13.2 billion light-years away. With its light waves redshifted—lengthened by the universal expansion of space—the galaxy was a scarlet blur whose glow took the Hubble eight days to gather. Because the Big Bang is thought to have occurred 13.7 billion years ago, the glow of galaxy UDFj-39546284 is a candidate for the oldest galactic light ever seen. But at the University of Arizona, Professor Dae Wook Kim dreams of probing deeper.

Growing up in South Korea, Kim was fascinated by physics, especially the enigma of light. "Light is something unique, special," the young professor tells me as we stand in the mirror lab gazing down at the huge white disc. "Light is the only thing humans can see with their own eyes but cannot touch. It has no volume or mass, except in the quantum realm—that's amazing—and it can't harm you except in special cases. Light is the fastest thing; it occupies no physical space, and it's almost free. That's why it has always been magical."

Before coming to the university in 2005, Kim had used telescopes but did not imagine building them. Yet he was captivated by these enormous mirrors and began working toward his Ph.D. by working with them. When his equations are exhausted, he compares the mirror lab to Michelangelo's *David.* To sculpt his masterpiece, Kim tells me, Michelangelo needed just three things—a chisel, his hands, and his genius. The tools fabricating these gigantic mirrors are the chisel. The sanders polishing the shiny disc below us are the project's hands. And the computer guiding it, nicknamed "the Matrix"—"I was a big fan of the movie," Kim says—provides the artist's genius.

Driven to make mirrors that find light from the first generation of galaxies, Kim wonders what that light might tell us. Though raised a Christian, the affable professor has no problem accomodating both science and Genesis. "The Bible is not my physics textbook," he says. Yet he retains the wonder of "Fiat lux." "It's taken four hundred years since the first telescope," Kim says, "but we are finally in touch with the other end. The fascinating thing is that we don't know what we're going to see. I think First

Light will be a lot different, something never imagined, never seen before. We are looking at our past. Light is telling us where we come from, letting us know where everything began. It will change human perception. It will change our understanding of the universe."

Chile's Giant Magellan Telescope is expected to begin its search for First Light in 2021. Kim stares down at the mirror and thinks of his family. "I have two sons, seven and six," he says. "Both born in Tucson. When these projects are done, my sons will be in college. Thanks to these telescopes, these mirrors, my boys will see to the end of the universe."

I left Arizona's College of Optical Sciences convinced that we have entered the Age of Light. The ages of steam and coal are long gone. With oil clinging to power, light is emerging as our deus ex machina. Light goes where nothing else can, gets there faster than anything else could, and brings back the images. If there are limits to light other than its cosmic speed limit, we have not tested them. If there is a final answer to the question "What is light?"—the answer Galileo dreamt of, the answer Einstein never stopped seeking—we have not found it. The search itself is eternal, spiraling back to the source.

Regardless of the weather, heedless of any clock, light comes late to Ireland on the shortest day of the year. All night the rain has fallen, not some drizzly Irish curtain but a pounding rain boding ill for hundreds of us who have come for the winter solstice sunrise at Newgrange. By seven A.M., almost two hours before first full light, the rain has let up but a dark, damp blanket envelops the Boyne River Valley an hour north of Dublin. Serpentine two-lane roads lead me through dimly lit villages toward my rendezvous with dawn. When I park in the Newgrange lot and join a few dozen hearty souls in the visitor's center, I hear more talk about Christmas than solstice. Late in December, this far north, light is more a yearning than either particle or wave, yet given the gloomy weather, few expect to see the sunrise that has drawn them from Australia, the United States, France, Russia, and throughout the British Isles. But if light is also hope, then it must spring eternal. Silently we board a bus and ride past black hedges, skirting lines of New Age revelers marching to Newgrange after a soggy overnight vigil at another sacred site nearby. Through fog and streetlights, we see their banners painted with swirling circles that mirror the petroglyphs carved here some five thousand years ago. Leaving the marchers behind, the bus takes us toward more clouds and only the slimmest chance of seeing the sun rise.

Older than Stonehenge and the pyramids, Newgrange is one of several passage tomb cemeteries built on the River Boyne in the fourth millennium BCE. Each primeval passage is aligned with the sun. A few miles away, the chamber at Dowth is aimed at the winter solstice sunset. The stone tunnels at nearby Knowth and Loughcrew run east–west to frame equinox sunrises in March and September. And here at Newgrange, a burial passage fifty feet long aligns perfectly with the sunrise on December 21. Since 3200 BCE, at eight fifty-four A.M. on the winter solstice, the rising sun has sent its light creeping along the subterranean stone path and into the burial chamber. Within minutes, shafts of sunlight illuminate tombs and urns, heralding the end of shortening days and endless nights that sap the soul. Today, the ancient chamber will once again light up on schedule—if clouds cooperate.

By eight A.M., some five hundred watchers are scattered across the dark, grassy slope facing southeast. Behind starry banners, marchers join hands and circle, chanting or singing. The rest of us stand, adjusting scarves and ski caps, shivering in the whipping winds. The horizon is gray as only an Irish horizon can be gray. At our backs stands Newgrange itself, an imposing cheesecake of white granite erected around the ancient entrance. The preservation was the work of archaeologist Michael J. O'Kelly. When he first came here in 1961, O'Kelly found a heap of stones and a circle of petroglyphs surrounding a hole in the ground. He considered Newgrange just another burial mound until locals began sharing stories of the solstice. Word had it that on December 21, deep inside the earth, the entire tomb lit up at dawn. O'Kelly dismissed the stories, thinking them somehow confused with Stonehenge. Then the ruddy, boisterous archaeologist decided to see for himself. Just before dawn on December 21, 1967, flashlight in hand, he found his way along damp walls to where the passageway opened to a twenty-foot ceiling. He sat and waited in the pitch-black.

The light inched up the tunnel in arrows. The first shafts struck the stone walls, then blazed along the floor—a beam at first, then a flare "lighting up everything," O'Kelly remembered, "until the whole chamber, side recesses, floor, and roof six meters above the floor were all clearly illuminated." Shivering, shaken, feeling akin to the ancients, O'Kelly quickly saw how neolithic technology had captured so much light. Above the entrance tunnel was an open square angled toward the rising sun. The radiance lasted seventeen minutes. When it dimmed, O'Kelly scrambled out and went to work. By 1982, with renovations finished, the crowds started to arrive. Solstice sunrise over Ireland's patchwork fields was lovely, but everyone wanted to be in the chamber and a waiting list soon formed. In 2000, when the wait reached ten years, the list became an annual lottery that now attracts thirty thousand entrants hoping to be chosen by local schoolchildren pulling tickets from a bowl. I was not chosen. I would have to wait outside to see the sun—perhaps.

At eight fifteen, a single drum stirs the circling dancers. The sky has begun to lighten and a half-moon skirts the clouds. The eastern horizon is still gray, but to the south, a clearing has appeared. Gale winds, along with numbing our cheeks, are pushing the front toward the Irish Sea. If clouds moving at twenty miles per hour can clear the path of the sun, approaching fifty times as fast, the miracle just might come off on schedule. The race is on. We stamp our feet, blow on our hands, and fix our eyes on the horizon. Hand-held light from smartphones reads 8:30. Beyond the river, above a scattering of cottages and sheep dotting the Kelly green, the sky turns a soft pink.

At eight forty-five, nine minutes before the moment, several dancers break free of their circle. Standing with feet apart, facing southeast, arms outstretched, they open themselves to ecstasy. Behind them, others steady their cameras or hold toddlers aloft, their cheeks brightening with the dawn. Along the horizon, clouds now tinged with red sweep eastward, leaving clear skies behind. But the pinpoint where the sky is brightest remains shrouded. The sun is losing this race. "What we need is a human sacrifice," someone with a brogue jokes, but all we can do is huddle and hope.

Moments later, beneath streaks of salmon overhead, the lottery winners line up to enter the chamber. Thinking themselves lucky, they duck through the low door. "Mind your head! Mind your head!" But we on the slope are more fortunate. Instead of single shafts, we stand dwarfed by the full sky, the moon, the horizon flooded now with blood oranges and reds, and a lone yellow dawn spot where the sun, shrouded or not, will rise any minute. Now the clouds are so bright we can't tell whether they have won the race. Now the eastern horizon is clear. The wind seems to have done its job, but where is the sun? Another minute, brighter, brighter. Morning spills across the fields and up the slope. Five hundred faces are turned toward the glare. A hillside holds its breath. Dancers stop. The drum stills. The whip of wind is the only sound. We wait. We wait.

And then as if the outcome were never in doubt, a blaze tops the edge of the earth. Cheers resound across the slope. Shouts and hurrahs. Arms extend and faces lift as if to inhale the light. A few revelers circle their fingers to hold the sun at arm's length while others hug or fall to their knees. Long shadows stretch up the grass to the white granite behind. Just as it did hours ago above the Ganges, moments ago above Stonehenge, light blesses again. Beside me, a little girl in her father's arms beams. "Oh, look at it!" she says, her eyes twinkling. "Look at it, look at it!" But we cannot look anymore. It is too bright. The full, blazing, blinding ball clears the horizon. We close our eyes and we can still see it.

Appendix

By the late twentieth century, light's new masters were bewildered. The laser technicians, the photonics engineers, the few who actually understood quantum optics, thought they had secularized light. Yet throughout the 1980s and 1990s, as photons were parsed into probability functions and density matrices, word spread of a light divine. Across America and around the world, reports came from those who had stood at the brink of death, peering, it seemed, into paradise. A boy in Melanesia, dying in his hut, saw a warm glow beckon him outside. When revived, the boy recounted how he had "walked along the beam of light, through the forest and along a narrow path." Elsewhere a Japanese man, recovered from a head-on collision, told of seeing a wall "made of golden light." Heart attack victims, their vital signs flatlined, came back to life with the most startling stories. Of hearing doctors declare them dead. Of seeing a tunnel, passing through it, meeting glowing figures on the other side. And always the light, warm, welcoming, "floating," "pure, crystal light," "so bright, so radiant," "not any kind of light you can describe on earth."

First it was just a few stories, then a few dozen, but finally came a deluge of stories about coming back from the brink of death. So similar were the reports that they earned an acronym, NDE, for "near-death experience." The term was coined by the psychiatrist who first catalogued the astonishing stories, Raymond Moody. In 1975, Moody's book *Life After Life* described fifty near-death experiences, each involving elements that are now fixtures in the NDE phenomenon/industry. The out-of-body experience. The tunnel. The light at its end. The emergence into a warm, nurturing glow, sometimes embodied in "a being of light." In an increasingly chaotic world, Moody had touched a nerve. By the end of the century, *Life After Life* had sold thirteen million copies and paradise was shining more brightly than it had since Dante.

Divine light had never disappeared—it had just been hidden under a bushel of skepticism. Throughout the nine centuries separating the light of Abbot Suger from the light of the first laser, photisms had been a periodic feature of spiritual life. Along with the visionaries—Joan of Arc, Teresa of Avila, Emanuel Swedenborg—ordinary people claimed to see holy light, which often heralded a vision of the Virgin Mary. In 1857, a glow appeared in a grotto outside Lourdes, France. Six decades later, in Fatima, Portugal, it was "white light gliding above the treetops." A half world away, the "Buddha lights" of Mount Wutai continued to sparkle, while a few Tibetan Buddhists saw rainbows ascending, the pure incarnation of their highest spiritual state: rainbow body. In India, practitioners of Kundalini yoga spoke of "liquid light" streaming through the spinal cord and into the brain. Eskimos described *quamaneg*, a bright light that let them see in the dark. Aboriginal shamans saw light crystals fall from heaven, while mourners in the British Isles swore that "corpse candles," blue lights along the ground, sometimes guided them to graves.

Nor was the light at the end of death's tunnel new. Ancient Greek sages described mystical lights that beckoned the dying. "The soul on the point of death," Plutarch wrote, "is met by a wonderful light and received into pure places and meadows with voices and dancing." The biblical stalwarts Jacob and Enoch, and later the Apostle Paul described visits to a floodlit paradise. Islam's prophet, Muhammad, journeyed around the world in a single night, stopping briefly in a radiant heaven. *The Tibetan Book of the Dead*, dating from the fourteenth century CE, described how the dying see the "Clear Light of Pure Reality."

Yet through the centuries, such reports became scarce, especially in the Western world. For all but the most devout, there was something suspicious about divine light, something fabricated, something Eastern. On into the 1970s, the skepticism continued. Those first brought back from the brink were reluctant to discuss what they had seen. "You learn very quickly that people don't take to this as easily as you would like them to," one man told Raymond Moody. "You simply don't jump on a little soapbox and go around telling everyone these things." Or at least, back then you didn't.

By the 1980s, however, the light of paradise shone forth unashamed. Attitudes toward death had softened, spirituality was diverging from organized religion, and medical advances were reviving more patients. With the endorsement of Dr. Elisabeth Kübler-Ross, who moved from defining the stages of death to collecting NDEs, the field of near-death studies was born. Soon, as if the Middle Ages had never ended, millions were seeing or seeking spiritual light. A 1982 Gallup Poll showed that one in six Americans had seen "the light" of an afterlife. And with reports surfacing in the media, even those who had never been to the brink knew of its light. Then, in 1990, in the blockbuster movie *Ghost*, a glowing Patrick Swayze, back from the beyond, kissed a radiant Demi Moore and the gates of heaven opened wide.

Scientists were intrigued but not inclined to go along. Dozens of studies narrowed the clinical sources of NDE light. At the University of Arizona, two psychologists, Willoughby Britton and Richard Bootzin, examined revived patients. Those who saw light, they hypothesized, would have unusually active right temporal lobes, the seat of epilepsy and dream states. As expected, they found such visionaries "physiologically distinct" from others—soldiers, auto accident victims, and heart patients—whose NDEs brought not rhapsody but terror. At the University of Kentucky, the neurologist Kevin Nelson found that NDE patients had often experienced REM sleep patterns while fully awake. The light of near death, Nelson argued, was "the light of REM consciousness and wakefulness blending into each other as death approached."

Other researchers found simpler explanations. Approaching death, the brain begins to shut down. The tunnel is not a tunnel but the world seen through retinas whose edges are deprived of blood. The light is not paradise but is sparked by chemicals emitted by a dying brainstem. Neurological causes explain why photisms are seen not just by the dying but by those under severe stress, including epileptics, women in childbirth, and fighter pilots under forces of several *g*'s. Examining the NDE literature, the neurologist Oliver Sacks saw classic examples of hallucinations. "Hallucinations," Sacks wrote, "cannot provide evidence for the existence of any metaphysical beings or places. They provide evidence only of the brain's power to create them."

As NDE stories spread, even Raymond Moody grew skeptical. Though predicting that further study will lead to "the greatest advances in the rational understanding of life after death since Plato," Moody frowned on "the avalanche of books on the subject . . . many that, to my personal knowledge, have been fabricated by unscrupulous self-promoters." Yet the avalanche continued, turning the light of heaven into a commodity. Since Moody's *Life After Life*, some two thousand books have taken readers down that well-trodden tunnel. Reports vary—slightly. Some tell of seeing Jesus, others of meeting God, a few of finding long-lost pets. But all have one element in common, as their titles suggest. "The Light Beyond" (1988); "Transformed by the Light" (1992); "Embraced by the Light" (1992); "Secrets of the Light" (2009); "Living in the Light" (2013): "Heading for the Light" (2014) . . .

NDEs raise more questions than they answer. If their light is a door to some beyond, why don't all revived patients have NDEs? (Less than 20 percent do.) If all NDEs have certain features, does that speak to the universality of an afterlife or to the universality of the human brain? Yet as sales pile up, the spate of books suggests a spiritual hunger as much as a hunger for easy money. Not everyone can get through Dante. Not all can afford a trip to Chartres or Saint-Denis. But for the price of a dinner out, anyone can buy a book and, suspending skepticism, see and feel a light that might await. In its fourth millennium, just as in its first, the Light Euphoric still beckons.

Acknowledgments

We may not all be children of light but we are all its disciples, and in undertaking this daunting subject, I was fortunate to meet many in its thrall. I am especially indebted to those who took the time to read portions of the manuscript and apply their own expertise. The list begins with my brother Doug, a mechanical engineer who helped me understand concepts ranging from blackbody radiation to whether Archimedes could really have torched ships with sunlight. My brother-in-law, David, applied the basics of his lifelong passion for chemistry to sign off on Daguerre's process. The artist Lynn Peterfreund offered boundless encouragement and made key suggestions on art history. And my wife, Julie, urged me not to be shy about stepping into the narrative. She also arranged with her colleague Don Smith, a former dean at the Cornell University College of Veterinary Medicine, to get the horse's eye that I sliced into, à la Descartes. Other general readers, chosen for their rare interest in everything, included Richard and Joan Godsey, Dick Gladden, Bill Adams, and Brooks Faris.

Many simply gawked when I told them what I was writing, but many more offered specific advice. The physicist Alan W. Hirshfeld of the University of Massachusetts at Dartmouth surveyed the book's scientific chapters and offered corrections and advice. The physicist Arthur Zajonc, author of *Catching the Light*, soothed my worries about quantum theory by pointing out how few people really understand it. My friend and Italian tutor, Nina Cannizzaro, shared her insights into Dante and the *Natural Magick* of her beloved Renaissance men. My friends Samar and Gabriel Moushabeck offered advice on Islamic science and culture. And though I have nothing but gratitude for librarians in general, I am specifically grateful to Mary Weidensaul, who, as I worked downstairs in the Jones Library in Amherst, Massachusetts, shared her enthusiasm for my search and brought me book after book. I am also indebted to Bard College's Language and Thinking Program, the annual freshman orientation known as L+T. It was after first teaching L+T's sweeping curriculum, from Aristotle to Richard Feynman, that I conceived the idea of gathering light from all disciplines. Several L+T colleagues, notably Thomas Bartscherer, Karen Gover, Kythe Heller, and Brian Schwartz, listened patiently to my neophyte's understanding of philosophy, religion, Islamic studies, and Einstein. And did not laugh.

Finally, I am indebted to my agent, Rick Balkin, for recognizing the potential in a proposal as amorphous as light. My editor at Bloomsbury, Jacqueline Johnson, handled this ethereal topic with sensitivity and patience. And where, I wonder, would I be without my dog, Jackson, with whom, like James Clerk Maxwell with his terrier, I shared my ideas as we walked through so many glorious sunrises.

Endnotes

Short titles are used for sources listed in the bibliography.

INTRODUCTION

"solar sponges": Zajonc, *Catching the Light*, 78.

"painting is light": Kriwaczek, *In Search of Zarathustra*, 104.

"If there is magic on this planet": Eiseley, *Immense Journey*, 15.

"When the great night comes": Cohen, *Chasing the Sun*, 213.

"children of light": Thessalonians 5:5.

"I am the light of the world": John 8:12.

"the Light of the heavens and the earth": Abdel Haleem, *The Qur'an*.

"the heaven of pure light": Mandelbaum, *Dante: Paradiso*, 266.

"light, seeking light": Shakespeare, *Complete Works*, 281.

"Hail, holy light": Milton, *Paradise Lost*.

"is reflected from the windows": Thoreau, *Walden*, 583.

CHAPTER I: "THIS LIGHT IS COME"

"this is the sun's birthday": E. E. Cummings, *Complete Poems*, 663.

"without form, and void": Genesis 1:2.

"darkness, blinding darkness": Walshe, *Long Discourses of the Buddha*, 410.

"darkness swathed in darkness": Sproul, *Primal Myths*, 183.

"darkness upon the face of the deep": Genesis 1:2.

"society's dream": Campbell and Moyers, *Power of Myth*, 48.

"the deepest and most important": von Franz, *Creation Myths*, 5, 9.

"all-spreading splendor": Long, *ALPHA*, 124.

"the fragments all grew lovely": Sproul, *Primal Myths*, 178.

"shouts and hurrahs": Hume, *Thirteen Principal Upanishads*, 214–15.

"God saw the light": Genesis 1:4.

"Before the beginning of time": Sproul, *Primal Myths*, 43.

"vomited up the sun": Leeming and Leeming, *Dictionary of Creation Myths*, 34.

"a cloud that floats in nothingness": Sproul, *Primal Myths*, 336.

"I am the one who openeth his eyes": Zajonc, *Catching the Light*, 39.

"When his eyes close, darkness falleth": Stewart, *Poetry and the Fate of the Senses*, 291.

"There's nothing but light": Sproul, *Primal Myths*, 333.

"What is spreading and covering the sky?": Ibid.

"Sun, sun, burn your wood": Nagar, *Indian Gods and Goddesses*, 117.

Kashi, "City of Light": Eck, *Banaras*, 10.

"darkness swathed in darkness": Zaehner, *Hindu Scriptures*, 10.

"the shining ones": Armstrong, *Case for God*, 12.

"This light is come": Griffith, *Complete Rig Veda*.

"Inconceivable, O Monks, is this Samsara": Sproul, *Primal Myths*, 194.

"There comes a time": Walshe, *Long Discourses of the Buddha*, 409.

"World of Radiance": Sproul, *Primal Myths*, 195.

"savory earth": Walshe, *Long Discourses of the Buddha*, 409.

"The result of this": Ibid., 410.

In the beginning: Genesis 1:1–3

"the two great lights": Genesis 1:16.

"let them be for signs": Genesis 1:14.

"An intimate connection exists": Dundes, *Sacred Narrative*, 195.

"not merely a story told": Ibid., 198–99.

"Was there a below?": Sproul, *Primal Myths*, 184.

CHAPTER 2: "THE THING YOU CALL LIGHT"

"You surely realize": Griffith, *Plato, The Republic*, 214.

"students of nature": Barnes, *Early Greek Philosophy*, 13.

"the cause of visible things": Bambrough, *Philosophy of Aristotle*, 55.

"the pure torch of the shining sun": Kirk, Raven, and Schofield, *Pre-Socratic Philosophers*, 258.

"luminous epiphanies of the gods": Pope, *The Iliad*.

"in a trail of light": Ibid.

"For the contemplation of the sun": Yonge, *Diogenes Laertius*.

"the great town of the tawny Acagras": Wright, *Empedocles*, 264.

"Alas, poor race of mortals": Barnes, *Early Greek Philosophy*, 197; Wright, *Empedocles*, 270.

"As when someone, intending a journey": Barnes, *Early Greek Philosophy*, 189–90.

"imprisoned in the membranes": Ibid.

"There are images or patterns": Park, *Fire Within the Eye*, 37.

"For external things": Lindberg, *Theories of Vision*, 2.

"And the pure fire": Jowett, *Plato: Complete Works*.

"You can look at the soul": Griffith, *Plato: The Republic*, 215.

"the lord of light": Jowett, *Portable Plato*, 550.

"Why should the eye": Park, *Fire Within the Eye*, 41.

"activity—the activity of what is transparent": Lindberg, *Theories of Vision*, 8.

"It is better": Beare, *Complete Aristotle*.

"see stars": Lindberg, *Theories of Vision*, 7.

"Vision occurs when the sensitive faculty": Ibid.

"The mote, which is seen": Ray, *History of Hindu Chemistry*, 10.

"Light is colored and illumines": Ibid., 7.

"one substance": Ibid., 8.

"Fire is both seen and felt": Ibid.

"watery": Ibid., 9.

"chief ingredient is light": Ibid.

"Shadows are two:" Graham and Sivin, "Systematic Approach to Mohist Optics," 120.

"universal love": Ibid., 105.

"understanding" or "intelligence": Nylan, "Beliefs About Seeing," 110.

"If a thing has no form": Ibid., 121.

"bright window dust": Teresi, *Lost Discoveries*, 201.

"That which lets now the dark": Capra, *Tao of Physics*, 106.

"The ray issues from the eye": Kheirandish, *Arabic Edition of Euclid's "Optics,"* 2.

"And that is what we wished to demonstrate": Ibid., 4.

"central fire": Teresi, *Lost Discoveries*, 205.

"evoke your astonishment": Toomer, *Diocles: On Burning Mirrors*, 40, 44.

"produce fire in temples": Ibid., 44.

"Archimedes' solar ray": "Mythbusters, Presidential Challenge."

"raypocalypse": Ibid.

"ought to appear fragmented": Smith, *Ptolemy's Theory of Visual Perception*, 92.

"heat in relation to the heater": Ibid., 78.

"visual flux": Ibid., 79.

"More comprehensive and technically demanding": Ibid., 50.

"Light is a 'nature'": Beare, *Complete Aristotle.*

"Let ABG be the arc": Smith, *Ptolemy's Theory of Visual Perception*, 170.

CHAPTER 3: "THE HIGHEST BLISS"

"In the paradise which you praise": Kriwaczek, *In Search of Zarathustra*, 105.

"beings of light": Assumen, *Manichaean Literature*, 52.

"numerous suns and full moons": "Sutra of Golden Light," Foundation for the Preservation of Mahayana Tradition, website (http://fpmt.org/education/teachings/sutras/golden-light-sutra/; accessed May 15, 2013).

"light from light": Pelikan, *Light of the World*, 65.

"a light to lighten": Luke 2:32.

"in thy light shall we see light": Psalms 36:9.

"God is Light": O'Collins, Gerald, and Meyers, *Light from Light*, 215.

"O Lord, master of yoga": Easwaran, *Bhagavad Gita.*

"If a thousand suns": Ibid.

"filled with amazement": Ibid.

"I see you, who are so difficult": Ibid.

"the gentle form of krishna": Ibid.

"The night is far spent": 1 Corinthians 13:12.

"Wise Lord": Fox, "Darkness and Light," 130.

"shine continually over the earth": Eliade, "Spirit, Light, and Seed," 14.

"the principles of Truth and Light": Irani, "The Gathas."

"Tell me truly, Ahura": Fox, "Darkness and Light," 131.

"best friend of the Lights": Assumen, *Manichaean Literature*, 11.

apostle of Jesus: Widengren, *Mani and Manichaeism*, 37.

"beloved of the beings of light": Rudolph, *Gnosis*, 337.

"Behold the illuminator of the hearts comes": Assumen, *Manichaean Literature*, 52.

"Manichaean": "Manichaean," A.Word.A.Day, Wordsmith.org. (http://wordsmith.org/words/manichean.html; accessed May 22, 2013).

"I thought that you, Lord God and Truth": Chadwick, *Augustine: Confessions.*

"lengthy fables": Ibid.

"insolent . . . wretched creatures": Hill, *Augustine: On Genesis*, 47.

"spiritual, not a bodily light": Ibid., 263.

"The soul of man": Chadwick, *Augustine: Confessions.*

"The highest bliss": Kapstein, *Presence of Light*, 286–87.

"the light that is most excellent": Griffith, *Complete Rig Veda.*

"Dawn on us with prosperity": Ibid.

"Suns blazing glory": "Sutra of Golden Light."

"a hundred thousand rays of light": "Larger Sutra of Immeasurable Life, Part 1."

"rainbow body": Kapstein, *Presence of Light*, 120.

"Buddha lights": Ibid., 205.

"Photisms bring a man": Eliade, "Spirit, Light, and Seed," 2.

"a Hebrew of the Hebrews": Philippians 3:5.

"yet breathing out threatenings": Acts 9:1.

"And it came to pass": Acts 22:6–9.

"There shall no man see me, and live": Exodus 33:20.

"dwell in the thick darkness": 1 Kings 8:12.

"To tell Paul that the Christ he knew": Buttrick, *Interpreter's Bible*, vol. 9, 327.

"a shining light": John 5:35.

"the true light": John 1:9

"a light into the world": John 12:46.

"the Light of the World": John 8:12, 9:5.

"unbegotten . . . eternal": Armstrong, *History of God*, 108.

"God, but not true God": Rubenstein, *When Jesus Became God*, 79.

"God from God, Light from Light": Pelikan, *Light of the World*, 67.

"unchanging and incorporeal light": O'Collins and Meyers, *Light from Light*, 197.

"His head and his hairs": Revelations 1:14–16.

"There shall be no night there": Revelations 22:5.

"brightness [that] comes": "Third Public Examination," St. Joan of Arc Center, website (www.stjoan-center.com/Trials/sec03.html; accessed May 10, 2015).

"not a brilliancy which dazzles": Lewis, *Life of St. Teresa of Jesus*.

"psychogenetic condition": Fox, *Spiritual Encounters with Unusual Light Phenomena*, 41.

"hysterics . . . should for even a short time": James, *Varieties of Religious Experience*, 252.

CHAPTER 4: "A GLASS LIKE A GLITTERING STAR"

"O God appoint for me": Zaleski and Zaleski, *Prayer: A History*, 59.

"place of light": Hillenbrand, *Islamic Architecture*, 132.

"While watching this spectacle": Behrens-Abouseif, *Minarets of Cairo*, 12–13.

"transparent stone": Smith, *Alhacen on Refraction, Vol. Two*, 244.

"God is the patron of the faithful": Qur'an 2:257.

"He will grant you a double share": Qur'an 57:28.

"Indeed the man from whom": Qur'an 24:40.

"God is the light of the heavens": Qur'an 24:35.

"His face shone like the sun": Matthew 17:2.

"the light of Tabor": O'Collins and Meyers, *Light from Light*, 134.

"uncreated light": Ibid., 22.

"from Alexandria to Baghdad": Wardrop, *Arabic Treasures of the British Library.*

"sacrifice and immolations": Toomer, *Diocles: On Burning Mirrors*, 36.

"if the perception of light": Singh, *Fundamentals of Optical Engineering*, 9.

"We ought not to be embarrassed": Jim Al-Khalili, *House of Wisdom*, 124.

pneuma: Lindberg, *Theories of Vision*, 11.

"luminous breath": Kheirandish, "Many Aspects of 'Appearances,'" 87.

"worthy of much derision": Lindberg, *Theories of Vision*, 25.

"Everything that has actual existence": Kheirandish, "The Many Aspects of 'Appearances,'" 19.

"It is manifest that everything": Lindberg, *Theories of Vision*, 19.

"I seek refuge in the Lord of Daybreak": Qur'an 113:1.

"a beam that is fairly wide": Sabra, *Optics of Ibn Al-Haytham*, 38.

"doctrines whose matter was sensible": Sabra, *Optics, Astronomy, and Logic*, 190.

"make himself an enemy": Al-Khalili, *House of Wisdom*, 152.

a "confusion": Sabra, "Ibn al-Haytham's Revolutionary Project in Optics," 90.

"eight digits long": Smith, *Alhacen on Refraction*, vol. 2, 277–78.

"Thanks be to God": Arafat and Winter, "Light of the Stars," 287.

"We find that when our sight": Smith, *Alhacen's Theory of Visual Perception, Vol. Two*, 343.

"quite impossible and quite absurd": Park, *Fire Within the Eye*, 78.

"crystalline humor": Sabra, *Optics, Astronomy, and Logic*, 192.

"For light creates beauty": Smith, *Alhacen's Theory of Visual Perception,Vol. Two*, 504.

"for in many visible forms": Ibid., 507.

"least light": Sabra, *Optics, Astronomy, and Logic*, 192.

"in cold storage": Smith, *Alhacen's Theory of Visual Perception, Vol. One*, xciv.

"the revolution in optics": Ibid., cxvii.

"May God protect us": Sabra, *Optics, Astronomy, and Logic*, 240.

"the effect of a light": Valkenberg, *Sharing Lights on the Way to God*, 229.

"The problems of physics": Sabra, *Optics, Astronomy, and Logic*, 239.

"beings of light": Eliade, *History of Religious Ideas*, 142.

soothing like "warm water": Razavi, *Suhrawardi and the School of Illumination*, 61.

CHAPTER 5: "BRIGHT IS THE NOBLE EDIFICE"

"grave inconveniences": Crosby et al., *Royal Abbey of Saint-Denis*, 15.

"moldering away": Panofsky, *Abbot Suger on the Abbey Church of St. Denis*, 6.

"for the sake of Christ": Ibid., 15.

"wonderful quarry": von Simson, *Gothic Cathedral*, 94.

"noble and common folk alike": Panofsky, *Abbot Suger on the Abbey Church of St. Denis*, 93.

"If, then, there exists": Fathers of the English Dominican Province, *St. Thomas Aquinas Summa Theologica*.

"the brightness of glory": Ibid.

"so living that it trembles": Mandelbaum, *Dante: Paradiso*, 16.

FOR BRIGHT IS THAT: Panofsky, *Abbot Suger on the Abbey Church of St. Denis*, 22.

"metaphysics of light": McEvoy, "Metaphysics of Light in the Middle Ages," 126–45.

"Divine Light": Pseudo-Dionysius the Areopagite, *De Coelesti Hierarchia*.

"the Divine Rays": Ibid.

"the First Gift and the First Light": Ibid.

"One Source of Light": Ibid.

"Every creature, visible or invisible": Panofsky, *Abbot Suger on the Abbey Church of St. Denis*, 20.

"unpolluted radiance of gold" and "the heavenly glow of silver": Moffitt, *Caravaggio in Context*, 31.

"The dull mind rises to the truth": Rudolph, *Artistic Change at Saint-Denis*, 53.

"The arrangement is almost too clever": Adams, *Mont St. Michel and Chartres*, 26.

"crown of light": Crosby et al., *Royal Abbey of Saint-Denis*, 23.

"most luminous windows": McDannell and Lang, *Heaven: A History*, 85.

"The church sparkles and gleams": Scott, *Gothic Enterprise*, 156.

"We, most miserable men": Panofsky, *Abbot Suger on the Abbey Church of St. Denis*, 15.

"It was as if they wished": Scott, *Gothic Enterprise*, 91.

"You paint for yourself": Rudolph, *Artistic Change at Saint-Denis*, 68.

called his style "modern": Scott Tiffany, director, "Building the Great Cathedrals," *Nova*, PBS, broadcast October 19, 2010.

"dechristianization": Schama, *Citizens*, 829.

"symphony in stone": Hugo, *Hunchback of Notre-Dame*, 107.

"vast poems" and "with exquisite joy": Auguste Rodin, "Gothic in the Cathedrals and Churches of France," 219–20.

"Chartres is no place for an atheist": Cecil Headlam, *Story of Chartres*, 2.

"brighten the minds": McDannell and Lang, *Heaven: A History*, 85.

"Physical light is the best": Ball, *Universe of Stone*, 243.

"the first bodily form": Riedl, *Robert Grosseteste on Light*, 10.

"rarified . . . until the nine heavenly spheres": Ibid., 14.

"expanding its outer parts": Ibid., 14–15.

"some kind of light": McEvoy, *Robert Grosseteste*, 91–92.

"we may make things": O'Collins and Meyers, 82.

"spent more than 2,000 pounds": Clegg, *First Scientist*, 37.

"It is possible that some other science": Burke, *Opus Majus of Roger Bacon, Volume II*, 420.

"a species . . . a propagation": Ibid., 490.

"it is impossible that there should be": Hyman and Walsh, *Philosophy in the Middle Ages*, 489.

"since any one has experience": Ibid.

"the wonders of refracted vision": Ibid., 582.

"that dumb ox": Cahn, *Classics of Western Philosophy*, 280.

"The dumb ox will yet make his lowing heard": Ibid.

"Objection 2": Aquinas, *Summa Theologica*.

"Whether light is a body?" Ibid.

"as soon as the sun": Ibid.

"an active quality consequent on": Ibid.

"fittingly assigned to": Ibid.

"water as crystal": Ibid.

"the created light is necessary": Ibid.

"There must be a limit": Mandelbaum, *Dante: Paradiso*, 16.

"although the light seen farthest off": Hollander and Hollander, *Dante: Paradiso*, 45.

"to limit some of that excessive splendor": Mandelbaum, *Dante: Purgatorio*, 128.

"where every light is muted": Mandelbaum, *Dante: Inferno*, 38.

"the lantern of the world": Mandelbaum, *Dante: Paradiso*, 4.

"like a diamond struck by sunlight": Hollander and Hollander, *Dante: Paradiso*, 41.

"like the fastest-flying sparks": Ibid., 177.

"as within a spark": Ibid., 201.

"We are all ready": Mandelbaum, *Dante: Paradiso*, 66.

"many living lights": Hollander and Hollander, *Dante: Paradiso*, 261.

"Now D, now I, now L . . .": Ibid., 491.

"Love Justice": Mandelbaum, *Dante: Paradiso*, 369.

"its color, gold": Ibid., 182.

"so many splendors": Hollander and Hollander, *Dante: Paradiso*, 571.

"a sun above a thousand lamps": Mandelbaum, *Dante: Paradiso*, 200.

"which shone more than a thousand miles": Ibid., 232.

"I saw a point": Hollander and Hollander, *Dante: Paradiso*, 762–763.

"higher heaven": Aquinas, *Summa Theologica*.

"light that flowed as flows a river": Hollander and Hollander, *Dante: Paradiso*, 817.

"The living ray that I endured": Ibid., 295.

"Here my exalted vision": Ibid., 917.

"figurando il Paradiso": Ibid., 626.

"representing paradise": Ibid., 627.

"picturing forth paradise": Binyon, *Portable Dante*, 487.

"describing paradise": Kline, "Paradiso, Cantos XXII–XXVIII."

"figuring of paradise": Cary, "Canto XXVIII."

"People can't seem to let go": Acocella, "What the Hell."

CHAPTER 6: *CHIARO E SCURO*

"portray faithfully all the visible works of nature": McMahon, *Treatise on Painting by Leonardo da Vinci*, 5.

"such an excellent imitator": Bondanella and Bondanella, *Vasari: Lives of the Artists*.

"indeed worthy of free minds": Grayson, *Alberti: On Painting and Sculpture*, 65.

"Painting possesses a truly divine power": Ibid., 61.

"painters should first of all": Ibid., 89.

"this little work of mine": Grafton, *Alberti: Master Builder*, 33.

"painter speaking to painters": Ibid., 55.

"Let me tell you": Ibid.

"a kind of sympathy among colors": Grayson, *Alberti: On Painting and Sculpture*, 93.

"the brightest gleams" and "the deepest shadows": Ibid., 91.

"those painters who use white": Ibid.

had done a "favor": Ibid., 107.

"and thereby proclaim": Ibid.

"the blood of physics": Barasch, *Light and Color*, 45.

"the Proof and Reason Why": Suh, *Leonardo's Notebooks*, 80.

"A shadow is never seen": Ibid., 81.

"You will note in drawing": Ibid., 82.

"is not its own color": Capra, *Leonardo*, 233.

"a stone flung into the water": Kemp, *Science of Art*, 47.

"O Painter": Filipczak, "New Light on Mona Lisa," 519.

"You imitator of nature": Kemp, *Science of Art*, 331.

"Look at the light": Richter, *Leonardo da Vinci: Notebooks*, 166.

"lighter the lower you depict it": Ibid., 96.

"When you represent in your work": Filipczak, "New Light on Mona Lisa," 551.

"O painter, have a court arranged": Ibid., 519.

"every color is more discernible": McMahon, *Treatise on Painting*, 83.

"color also dissolves progressively": Grayson, *Alberti: On Painting and Sculpture*, 91.

"paint so that a smoky contour": Barasch, *Light and Color*, 74.

"Tell me if anything was ever done": Richter, *Leonardo da Vinci: Notebooks*, 20.

"Whoever reads these notes": McMahon, *Treatise on Painting*, xxi.

"brief rules regarding lights": Ibid., 286.

"the image of the divine mind": Haydocke, *Tracte Containing the Arte of Curious Paintings*, 135.

"has so great force": Kemp, *Science of Art*, 270.

"that which the sunbeams make": Haydocke, *Tracte Containing the Arte of Curious Paintings*, 175.

"No one who loves Caravaggio": Hibbard, *Caravaggio*, 260.

"The light and darkness of the world": Robb, *Man Who Became Caravaggio*, 128.

"too much light": McMahon, *Treatise on Painting*, 259.

"always more powerful than light": Ibid., 261.

"A certain Michael Angelo van Caravaggio": Moffitt, *Caravaggio in Context*, 201–202.

"The painter Caravaggio has left Rome": Ibid., 6.

"the optical look": Hockney, *Secret Knowledge*, 51.

"The camera obscura served Vermeer": Ibid., 251–52.

"a master of illusion": Liedtke, *View of Delft*, 249.

"Such an invention": Kemp, *Science of Art*, 163.

"is almost entirely lacking" and "tends to be inconclusive": Ibid., 193.

"is not to diminish": Hockney, *Secret Knowledge*, 14.

"only saying that artists": Ibid., 13.

"hang this piece in a strong light": Van de Wetering, *Rembrandt*, 251.

"in a stroke of daring": Schama, *Rembrandt's Eyes*, 422.

"his own sun-god": Wallace, *World of Rembrandt*, 110.

"I should be happy to give ten years": Van de Wetering, *Rembrandt*, 155.

CHAPTER 7: "INVESTIGATE WITH ME WHAT LIGHT IS"

"age of the marvelous": Moffitt, *Caravaggio in Context*, 190.

"the shutting in": Ekrich, *At Day's Close*, xxxii.

"The night has fallen": Koslofsky, *Evening's Empire*, 12.

"Who goes out at night": Ibid., 43.

"natural magick": Della Porta, *Natural Magick*.

"when someone looked in the mirror": Reeves, *Galileo's Glasswork*, 25.

"all ships passing near the column": Ibid., 34.

"mirrors of fire": Capra, *Leonardo*, 220.

"Changing Metals . . . Counterfeiting Gold": Della Porta, *Natural Magick*, 1.

"constrains women to cast off their clothes": Burnett, *Descartes and the Hyperbolic Quest*, 7.
"for ten, twenty, a hundred": Della Porta, *Natural Magick*, 371.
"an instrument for seeing far": Reeves, *Galileo's Glasswork*, 72.
"a friend a thousand miles distant": Ibid., 73.
"With a Concave you shall see": Park, *Fire Within the Eye*, 368.
"step-daughter of astronomy": Koestler, *The Sleepwalkers*, 245.
"chronic putrid wounds in my feet": Ibid., 233.
"something akin to the soul": Ferris, *Coming of Age in the Milky Way*, 77.
"For when the most wise founder": Kepler, *Optics*, 1.
"the most excellent thing": Ibid.
"poured about the moon": Ibid., 302.
"a looking glass of Astrology": Reeves, *Galileo's Glasswork*, 117.
"light held together by moisture": Ferris, *Coming of Age in the Milky Way*, 95.
"the universal starting point": Ibid.
"seized with a desire": Reston, *Galileo: A Life*, 86.
"Having dismissed earthly things": Van Helden, *Galileo Galilei*, 38.
"For the galaxy is nothing else": Ibid., 62.
"that He has been pleased": Reston, *Galileo: A Life*, 99.
"O telescope, instrument of much knowledge": Ferris, *Coming of Age in the Milky Way*, 95.
"it would be of": Frova, Marenzana, and McManus, *Thus Spoke Galileo*, 192.
"when their ultimate and highest resolution": Ibid., 413.
"the opposite light was instantaneous": Ibid., 427.
"What a sea we are gradually slipping into": Ibid.
"I had always felt so unable": Frova, Marenzana, and McManus, 414n.
"rose with the sun": Beare, *Complete Aristotle*.
"dazzle mine eyes": Shakespeare, *Complete Works*, 99.
"a firm and steadfast resolution": Clarke, *Rene Descartes*, 16.
"the great book of the world": Ibid., 10.
"Just as painters": Gaukroger, *Descartes: An Intellectual Biography*, 221.
"investigate with me what light is": Clarke, *René Descartes*, 87.
"very violent motion": Ibid.
"The same motion which is in the flame": Ibid., 89.
"my first master in optics": Aczel, *Descartes's Secret Notebook*, 63.
"in an instant": Mahoney, *Descartes: Le Monde*, 173.
"plenum": Zajonc, *Catching the Light*, 92.
"It has sometimes doubtless happened to you": Mahoney, *Descartes: Le Monde*, 67.

"In the same way": Ibid., 69.

"never again shall there be a flood": Genesis 9:11

"wondrous bow": Pope, *The Iliad.*

"a reflection of sight to the sun": Beare, *Complete Aristotle.*

"The harder and firmer": Shapiro, "Study of the Wave Theory of Light," 156.

"I do not believe it is necessary": Sabra, *Optics, Astronomy, and Logic*, 361.

"principle of least time": Zajonc, *Catching the Light*, 287.

"The most distant peoples should come": Koslofsky, *Evening's Empire*, 16.

"Before this age": Ibid., 259.

CHAPTER 8: "IN MY DARKEN'D CHAMBER"

"I took a bodkin": White, *Isaac Newton*, 61.

"a motion of spirits": Gleick, *Isaac Newton*, 76.

"Des-cartes" or just "Cartes": Hall, "Sir Isaac Newton's Notebook, 1661–1665," 244.

"a man going or running": Gleick, *Isaac Newton*, 44.

"A little fellow": Ibid., 29.

"Punching my sister": Christianson, *In the Presence of the Creator*, 199.

"blew": Hall, "Sir Isaac Newton's Notebook, 1661–1665," 198.

"Having darken'd my chamber": Turnbull, *Correspondence of Isaac Newton, 1661–1675*, 92.

"a disproportion so extravagant": Ibid.

"oil olive": Newton, *Opticks.*

"I could observe no such curvity in them": Turnbull, *Correspondence of Isaac Newton, 1661–1675*, 94.

"my darken'd chamber": Newton, *Opticks.*

"What is redness?": Park, *Fire Within the Eye*, 37.

"flames which emanate": Jowett, *Portable Plato.*

"Have no need of color": Stallings, *Lucretius.*

"has no cause other than that shame": Smith, *Alhacen's Theory of Visual Perception, Vol. Two*, 442.

"appear varied according to the lights": Hills, *Light of Early Italian Painting*, 67.

"imagination and fantasy and invention": Gleick, 78.

"no sensible changes": Newton, *Opticks.*

"Red and blue make purple": Hall, "Sir Isaac Newton's Notebook, 1661–1665," 248.

"Hence redness, yellowness &c": Ibid.

"The recent invention of telescopes": Shapiro, *Optical Papers of Isaac Newton, Volume 1*, 47.

"It might perhaps seem a vain endeavor": Ibid.

"lest you think": Ibid., 51.

"to the walls": Hall, *All Was Light*, 46.

"the prejudic'd and censorious multitude": Turnbull, *Correspondence of Isaac Newton, 1661–1675*, 92.

"As the rays of light differ": Christianson, *In the Presence of the Creator*, 150.

"But the most surprising and wonderful composition": Ibid., 151.

"an impression on the retina": Gleick, *Isaac Newton*, 80.

"the Excellent Discourse of Mr. Newton": White, *Isaac Newton*, 177.

"Even those very experiments": Christianson, *In the Presence of the Creator*, 156–57.

"Colour is nothing but the Disturbance": Tumbull, *Correspondence of Isaac Newton, 1661–1675*, 110.

"cause no determinate": Ibid., 145.

"Neither do I see": Christianson, *In the Presence of the Creator*, 179.

"voyaging through strange seas": Gleick, *Isaac Newton*, 196.

"In hunting for a shadow hitherto": Westfall, *Never at Rest*, 245.

God's "inner light": *Encyclopedia Brittanica Online*, s.v. "Inner Light" (www.britannica.com/EBchecked/topic/288537/Inner-Light; accessed December 1, 2014).

"Let us be honest": Barocas, *Nature of Light*, 125.

"Shadow is divided into two parts": McMahon, *Treatise on Painting*, 210.

"is propagated or diffused": Park, *Fire Within the Eye*, 190.

"The modification of light": Barocas, *Nature of Light*, 144–45.

"HE LIVED AMONG US": Park, *Fire Within the Eye*, 193.

"We will cease to be astonished": Barocas, *Nature of Light*, 201.

"Were I to assume an Hypothesis": Turnbull, *Correspondence of Isaac Newton, 1661–1675*, 363.

"show'd that Mr. Newton had taken": Christianson, *In the Presence of the Creator*, 193.

"many slender Arcs of Colours": Newton, *Opticks*.

"I suppose he will allow me": Hall, *All Was Light*, 71.

"according to Mr. Newton's directions": Christianson, *In the Presence of the Creator*, 197.

"for I see a man must either": Ibid., 198.

"Does he eat & drink": Gleick, *Isaac Newton*, 162–63.

"for having been at pains": White, *Isaac Newton*, 223.

"You say, you dare not yet": Gleick, *Isaac Newton*, 178.

"the most fearful, cautious, and suspicious": Dolnick, *Clockwork Universe*, 5.

"fits of easy Reflexion" or "fits of easy transmission": Newton, *Opticks*.

"stifled or lost": Hall, *All Was Light*, 120.

"foreign light": Newton, *Opticks*.

"vulgar": Ibid.

"an Introduction to Readers": Ibid.

"Grimaldo": Ibid.

"like the Tails of Comets": Ibid.

"Much subtiler Medium than Air": Ibid.

"Are not the Rays of Light in passing": Ibid.

"Are not the Rays of Light very small Bodies": Ibid.

"why may not Nature change Bodies": Ibid.

"the Worship of false Gods": Ibid.

"all was light": Pope, *Complete Works.*

"The book makes no Noyse": Hall, *All was Light*, 180.

"I have seen a professor of mathematics": Dolnick, *Clockwork Universe*, 45.

"whose equal is hardly found": Morley, *Letters on England by Voltaire.*

"with more dexterity": Ibid.

"fire itself thrown from the Sun": Hanna, *Elements of Sir Isaac Newton's Philosophy*, 14–15.

"Is it because they are born in France": Ibid., 103.

"All Paris resounds with Newton": Zajonc, *Catching the Light*, 87.

"There is a new universe opening": Hall, *All Was Light*, 226.

"And then I stay'd till by the Motion": Newton, *Opticks.*

CHAPTER 9: "A WILD AND HARMONIZED TUNE"

"meddling intellect": Hirsh, *World of Turner*, 58.

"the sublime": Blake, *Complete Works.*

"our philosophic sun": Nicholson, *Newton Demands the Muse*, 38

"the godlike man": Ibid., 39.

"who was himself the light": Ibid., 43.

"With what astonishment are we transported": Hastie, *Kant's Cosmogony.*

"the chaotic material of future suns": Ferris, *Coming of Age in the Milky Way*, 157.

"Never was I so devout": Landon, *Haydn: Chronicle and Work*, 116.

"The most profound silence": Geiringer, *Haydn: A Creative Life in Music*, 145.

"*Im Anfange schuf*": Haydn, *Die Schöpfung.*

"When light broke out": Jager, *Book of God*, 22.

"a pearl through a burnt glass": Hirsh, *World of Turner*, 54.

"the sun is God": Birch, *Ruskin on Turner*, 81.

"It seemed to amuse him": "Death of J. M. W. Turner, Esq., R.A.," *London Times*, December 23, 1851.

"The man must be loved for his works": Shanes, *Life and Masterworks of J. M. W. Turner*, 28.

"sister arts": Shanes, *Turner: The Great Watercolours*, 16.

"Why are you crying like that": Hamilton, *Turner: A Biography*, 61.

Light, Primary Light: Shanes, *Turner's Human Landscape*, 279.

"Sky and water. Are they not glorious?": Birch, *Ruskin on Turner*, 99.

"Dutch light": DeKroon, *Dutch Light*.

"delights to go back": Stewart, *Poetry and the Fate of the Senses*, 291.

"tinted steam": Hirsh, *World of Turner*, 40.

"I am nature": Jackson Pollock, "One: Number 31, 1950," Museum of Modern Art website, "The Collection" (www.moma.org/collection/object.php?object_id=78386; accessed May 10, 2015).

"A Typhoon bursting in a Simoom": Hirsh, *World of Turner*, 169.

"Why then do you blame Turner": Birch, *Ruskin on Turner*, 39.

"Reason and Newton, they are quite two things": Blake, *Complete Works*.

"Newtonian phantasm": Ibid.

"to a prism": Nicholson, *Newton Demands the Muse*, 1.

"to Newton's health": Ibid.

"soulless fireball": Ferber, *Romanticism*, 89.

"if there are in this world": Matthaei, *Goethe's Color Theory*, 202.

"reserve the right": Ibid.

"when it came to color": Ibid., 198.

"Like the entire world": Ibid., 199.

"I immediately spoke out loud": Ibid.

"I have yet to find someone": Ibid., 30.

"delicate empiricism": Seamon and Zajonc, *Goethe's Way of Science*, 2.

"It is a calamity": Ibid.

"not so much a study of color": Williams, *Life of Goethe*, 263.

"does not see a pure phenomenon": Matthaei, *Goethe's Color Theory*, 65.

"There arises no prismatic color": Sepper, *Goethe Contra Newton*, 143.

"the simplest, most homogenous": Steiner, *Goethe's Conception of the World*, 160.

"must move anyone who is not depraved": Sepper, *Goethe Contra Newton*, 143.

"a serene, gay, softly exciting character": Matthaei, *Goethe's Color Theory*, 169.

"a kind of contradiction": Ibid., 170.

"conveys an impression of gravity": Ibid., 172.

"The eye experiences": Ibid., 174.

"Lively nations": Ibid., 179–180.

"the female sex in youth": Ibid., 180.

"bons vivants," "orators, Historians, Teachers," "lovers and poets": Ibid., 189.

"guilds": Sepper, *Goethe Contra Newton*, 183.

"I have quarreled": Ibid., 181.

"What? Light should only exist": Zajonc, *Catching the Light*, 340.

"Color is the soul of nature": "Color in the Waldorf School: Van James," Waldorf Today, website (www.waldorftoday.com/2010/12/color-in-the-waldorf-school-van-james/; accessed May 10, 2015).

Selene (Greek), Soma (Indian): Cashford, *Moon: Myth and Image*, 25, 233, 234.

"the Mansion of the Moon": Griffith, *Complete Rig Veda*.

"gentle moon": Barnes, 182.

"air cut off by the fire": Wright, 25.

"O, swear not by the moon": Shakespeare, *Complete Works*, 345.

"across the heath": Attlee, *Nocturne*, 241.

"gracious moon": Ibid., 172.

"The moon was to me dear": Wordsworth, *Complete Poetical Works*.

"that orbed maiden": Shelley, *Complete Poetical Works*.

"passing strange and wonderful" and "believe I worship the moon": Holmes, *Shelley: The Pursuit*, 630.

"like a dying lady": Ibid., 611.

"splendid moon," "glowing moon": Keats, *Delphi Complete Works*.

"What is there in thee, Moon": Ibid.

"They look'd up to the sky": Byron, *Don Juan*.

"And should he have forgotten": Ibid.

"tender moonlit situation": Ibid.

"But lover, poet, or astronomer": Ibid.

"a striking example": Sepper, *Goethe Contra Newton*, 201n.

"a third-rate oratorio": Geiringer, *Haydn*, 307.

"more light": Orlet, "Famous Last Words." *Vocabula Review*, online magazine, July–August 2002 (www.vocabula.com/index.asp; accessed February 24, 2014).

CHAPTER 10: UNDULATIONS

"I must own I am much in the dark about light": Weiss, *Brief History of Light*, 24.

"whilst the great ocean of truth": Gleick, *Isaac Newton*, 19.

"calculating machine": Maxwell, *Five of Maxwell's Papers*.

"Because it's so much fun": Ferris, *Coming of Age in the Milky Way*, 180.

"that strange substance": Newton, *Opticks*.

"Every Ray of Light": Ibid.

"Amongst transparent bodies": Huygens, *Treatise on Light*.

"marvelous": Ibid.

"the ordinary" and "the extraordinary": Ibid.

"It seems as though": Ibid.

"great power of purifying": Rabinowitch, "An Unfolding Discovery," 2875–76.

"Phenomenon Young": Robinson, *Last Man Who Knew Everything*, 62.

"one of the most acute men": Ibid., 7.

"may be said to have been": Ibid., 19.

"one or two difficulties": Ibid., 103.

"fits of easy Reflexion" or "fits of easy transmission": Newton, *Opticks*.

"the white heart of a wind furnace": Robinson, *Last Man Who Knew Everything*, 103.

"has converted that prepossession": Young, "Classics of Science," 273.

"Suppose a number of equal waves": Robinson, *Last Man Who Knew Everything*, 107.

"Now I maintain that similar effects": Ibid.

"perhaps the single most influential experiment": Mollon, "Origin of the Concept of Interference," 814.

"We may be allowed to infer": Crew, *Wave Theory of Light*, 74.

"one of the most incomprehensible suppositions": Robinson, *Last Man Who Knew Everything*, 116.

"Much as I venerate the name of Newton": Barocas, *Nature of Light*, 240.

"Let him make the experiment": Robinson, *Last Man Who Knew Everything*, 117.

"striking resemblance": Ibid., 159.

"The ray issues from the eye": Kheirandish, *Arabic edition of Euclid's Optics*, 2.

"ray of certain intelligible Light": Park, *Fire Within the Eye*, 89.

"with motions like that of an Eel": Newton, *Opticks*.

"but Light is never shown to follow": Ibid.

"a professor only in name": Levitt, "Editing Out Caloric," 54.

"the tumult of carnage": Buchwald, *Rise of the Wave Theory of Light*, 24.

"I pass whole days": Arago, *Biographies of Distinguished Scientific Men*, 165.

"the cosine squared": Buchwald, *Rise of the Wave Theory of Light*, 47.

"really French": Aczel, *Pendulum*, 67.

"I've about broken my head": Buchwald, *Rise of the Wave Theory of Light*, 114.

"I tell you I am strongly tempted": Ibid., 116.

"a taste for exactitude": Levitt, *A Short Bright Flash*, 87.

"Using a lens of 2 mm. focus": Crew, *Wave Theory of Light*, 82.

"as much as the state of the sky permitted": Buchwald, *Rise of the Wave Theory of Light*, 137.

"serve to prove the truth": Barocas, *Nature of Light*, 246.

"year without a summer": Evans, "Blast from the Past."

"I have decided to remain": Levitt, *A Short Bright Flash*, 43.

"to determine by precise experiments": Barocas, *Nature of Light*, 250.
"Luminous *rays*": Buchwald, *Rise of the Wave Theory of Light*, 170.
"Light from which the *rays* emanate": Ibid.
"The motions of the *rays*": Ibid.
"hard battle" led by "General Arago": Ibid., 169.
"One can add a motion of rotation": Ibid., 172.
"does not dread difficulties of analysis": Ibid., 188.
"the molecules were in their equilibrium positions": Ibid., 196–197.
"Nature simple and fertile": Levitt, *A Short Bright Flash*, 46.
"elementary rays": Levitt, "Editing Out Caloric," 62.

CHAPTER II: *LUMIÈRE*

"the sky had become a grating": Hugo, *Les Misérables*, 175.
"the Versailles of the sea": Levitt, *A Short Bright Flas*h, 67.
"lenses by steps": Ibid., 58.
"like a star of the first magnitude": Ibid., 92.
"lantern of fear": Petroski, *Success Through Failure*, 17.
"the most astonishing thing": Gernsheim and Gernsheim, *L. J. M. Daguerre*, 9.
"Palace of Light": Lowry and Lowry, *Silver Canvas*, 4.
"marvel by M. Daguerre": Gernsheim and Gernsheim, *L. J. M. Daguerre*, 35.
"The most striking effect": Ibid., 15.
***tableau magique*:** Le Gall, *La Peinture Mecanique*, 28.
"a veritable triumph": Gernsheim and Gernsheim, *L. J. M. Daguerre*, 16.
"An epoch in the history of painting": Ibid., 17.
"Entitles M. Daguerre": Ibid., 2.
"Papa, is the goat real?": Ibid., 30.
"Won't one ever succeed": Ibid., 47.
"lunar silver": Daguerre, *History and Practice of Photogenic Drawing*, 20.
"photograms": Rudnick, "The Photogram—A History."
"Bologna stone": Gernsheim and Gernsheim, *L. J. M. Daguerre*, 58.
"I have seized the fleeting light!": Daguerre, *History and Practice of Photogenic Drawing*, 20.
"He is always at the thought": Gernsheim and Gernsheim, *L. J. M. Daguerre*, 1.
"It is said that M. Daguerre": Ibid., 71.
"The wish to capture evanescent reflections": Ibid., 2.
"The daguerreotype is not merely an instrument": Ibid., 78.
"the eye and pencil": Daguerre, *History and Practice of Photogenic Drawing*, 39.

"Silver iodide!"; "Quicksilver!"; "Hypo-sulfite of soda!": Gernsheim and Gernsheim, *L. J. M. Daguerre*, 98–99.

"daguerreotypomania": Ibid., 103.

"Everyone wanted to copy the view": Ibid., 100.

"Rembrandt perfected": Ibid., 87.

"painting is dead": Schneider, *World of Manet*, 92.

"a great manufactory of putrefaction": Jones, *Paris*, 301.

"The motif for me is nothing": Pissarro, *Monet's Cathedral*, 21.

"Drunk on sunshine": Dustan, *Painting Methods of the Impressionists*, 63.

"The Impressionists": Callen, *Techniques of the Impressionists*, 58.

"Black is not a color": Schneider, *World of Manet*, 105.

"brown gravy": Ibid., 22.

"Their aim was to perceive": Callen, *Techniques of the Impressionists*, 66–67.

"People entered it": Schneider, *World of Manet*, 27.

"the ultimate in ugliness": Ibid., 57.

"Proceeding from intuition to intuition": Roque, "Chevreul and Impressionism," 28.

"Imagine that I get up": Pissarro, *Monet's Cathedral*, 18.

"heartbreaking": Ibid., 19.

"My goodness, they are not very far-sighted": Ibid., 20.

"no longer the oblique light": Ibid., 10.

"the last thing": Ibid., 34.

"Precisely by slicing out this moment": Sontag, *On Photography*, 15.

CHAPTER 12: "LITTLE GLOBE OF SUNSHINE"

"This crystal tube the electric": Maxwell, "To the Chief Musician upon Nabla."

"a certain most subtle Spirit": Newton, *General Scholium*.

"A new sort of urban star": Stevenson, "A Plea for Gas Lamps," 280.

"plea for gas lamps": Ibid.

"sober scientific electricians": "Electric Illumination," *New York Times*, June 15, 1879, 5.

"philosophers": Faraday, *Chemical History of a Candle*.

"as fashionable an amusement as the Opera": Hamilton, *A Life of Discovery*, 342.

"seemed to carry him": Ibid., 343–44.

"all phenomena are produced by the same power": Williams, *Michael Faraday*, 138.

"out of hieroglyphics": Sharlin, *Making of the Electrical Age*, 57.

"want of mathematical knowledge": Forbes and Mahon, *Faraday, Maxwell and the Electromagnetic Field*.

"imponderable fluids": Williams, *Michael Faraday*, 167.

"lines of force": Ibid., 387.

"most probably, to light": Forbes and Mahon, *Faraday, Maxwell and the Electromagnetic Field*.

"Only the very strongest conviction": Williams, *Michael Faraday*, 385.

"succeeded in . . . magnetizing": Faraday, "Experimental Researches in Electricity," 2.

"will most likely prove": Williams, *Michael Faraday*, 386.

"ALL THIS IS A DREAM": Hamilton, *A Life of Discovery*, 334.

"I was at first almost frightened": Sharlin, *Making of the Electrical Age*, 80.

"We come here to be philosophers": Faraday, *Chemistry of a Candle*.

"What's the go o' that?": Forbes and Mahon, *Faraday, Maxwell and the Electromagnetic Field*.

"It's the sun!": Tolstoy, *James Clerk Maxwell*, 12.

"Gin a body": Forbes and Mahon, *Faraday, Maxwell and the Electromagnetic Field*.

"the department of the mind": Ibid.

"flight of brick-bats": Campbell, *Life of James Clerk Maxwell*, 169.

"a calculating machine": Maxwell, *Five of Maxwell's Papers*.

"feel the electrical state": Forbes and Mahon, *Faraday, Maxwell and the Electromagnetic Field*.

"I also have a paper afloat": Tolstoy, *James Clerk Maxwell*, 126.

"One scientific epoch ended": Forbes and Mahon, *Faraday, Maxwell and the Electromagnetic Field*.

"an aethereal medium": Maxwell, "A Dynamical Theory of the Electromagnetic Field," 464.

"This velocity is so nearly that of light": Ibid., 466.

"Calorific Rays": Wilk, *How the Ray Gun Got Its Zap*, 200.

"mutual embrace": Bodanis, *E=mc2*, 47.

"lapsed into mysticism": Forbes and Mahon, *Faraday, Maxwell and the Electromagnetic Field*.

"I tried the full power": Wrege and Greenwood, "William E. Sawyer," 34.

"sufficiently satisfactory, when looked at": Jonnes, *Empires of Light*, 55.

"shall be compact and of small size": U.S. Patent 208,252 A.

"too many unsolved technical problems": Maxim, *A Genius in the Family*, 90.

"It was all before me": Brox, *Brilliant*, 111.

"There was the light": Jonnes, *Empires of Light*, 57.

his "muckers": Collins and Gitelman, *Thomas Edison and Modern America*, 16.

"a minor invention every ten days": Millard, *Edison and the Business of Invention*, 6.

"the entire lower part of New York City": Jonnes, *Empires of Light*, 56.

"The more resistance": Brox, *Brilliant*, 115.

"T.A.", short for "try again": Jehl, *Menlo Park Reminiscences*, 338.

"I have carbonized and used": U.S. Patent 223,898.

"cities and villages lighted": "Electric Illumination," *New York Times*, June 15, 1879, 5.

"some very interesting experiments": Collins and Gitelman, *Thomas Edison and Modern America*, 97.

ELECTRIC ILLUMINATION: Jonnes, *Empires of Light*, 65.

"Edison's electric light, incredible as it may appear": Ibid.

"a little globe of sunshine": McClure, *Edison and His Inventions*, 159.

"strange weird light": Ibid., 23.

"The people, almost with bated breath": Ibid.

"One hundred dollars they don't go on!": Jonnes, *Empires of Light*, 84.

"It was not until about 7 o'clock": "Edison's Electric Light," *New York Times*, September 5, 1882, 8.

"war of the currents": Brox, *Brilliant*, 125.

"all I'll ever fool with": Ibid.

"Just as certain as death": Jonnes, *Brilliant*, 138.

"a city of Living Light": Freeberg, *Age of Edison*, 231.

"It is the most economical artificial light": Collins and Gitelman, *Thomas Edison and Modern America*, 125–27.

"Any sufficiently advanced technology": Kaku, *Physics of the Impossible*, 3.

CHAPTER 13: *C*—EINSTEIN AND THE QUANTA, PARTICLE, AND WAVE

"The more important fundamental laws": Swenson, *Ethereal Aether*, 125.

"laughing philosopher": Isaacson, *Einstein*, 28.

"a positively fanatic orgy of free thinking": Einstein, *Autobiographical Notes*, 3–5.

"If a person could run after a light wave": Isaacson, *Einstein*, 26.

"such a thing is impossible": Ibid.

"disturb the universe": Eliot, *Complete Poems and Plays*, 5.

"not a movement": Beare, *Complete Aristotle*.

"is emitted with infinite speed": Park, *Fire Within the Eye*, 62.

"stellar aberration": Ibid., 228.

"sufficiently quiet": Michelson, "Experimental Determination of the Velocity of Light."

"the exquisite gradations": Michelson, *Light Waves and Their Uses*, 1–2.

"the aether the stars graze upon": Stallings, *Lucretius*.

"solary fewell . . . rarer, subtiler": Christianson, *In the Presence of the Creator*, 89.

"There can be no doubt": Park, *Fire Within the Eye*, 286.

"ether wind": Gardner, *Relativity Simply Explained*, 18.

"ether drag": Livingston, *Master of Light*, 70.

"The second swimmer will always win": Ibid., 77.

"My Dear Mr. Bell": Swenson, *Ethereal Aether*, 69.

"One thing we are sure of": Ibid., 77–78.

"to a task he felt": Livingston, *Master of Light*, 112.

"softening of the brain": Ibid.

"most famous failed experiment": American Physical Society, "Michelson and Morley."

"If there be any relative motion": Swenson, *Ethereal Aether*, 95.

"Remove from the world the luminiferous ether": Park, *Fire Within the Eye*, 287.

"If I pursue a beam of light": Isaacson, *Einstein*, 26.

"psychic tension": Overbye, *Einstein in Love*, 131.

"the light medium": Isaacson, *Einstein*, 117.

"The difficulty to be overcome": Overbye, *Einstein in Love*, 133.

"patent slave": Kumar, *Quantum*, 31.

"a storm broke loose in my mind": Ibid., 33.

"No one is allowed to predicate things": Overbye, *Einstein in Love*, 100.

"very revolutionary": Ibid., 123.

"ultraviolet catastrophe": Kumar, *Quantum*, 47.

"For six years I had struggled": Lisa Randall, *Warped Passages: Unraveling the Mysteries of the Universe's Hidden Dimensions*, (New York: HarperCollins, 2005), 123.

"an act of desperation": Ibid.

quantus, **"how much?":** Robert P. Crease and Charles C. Mann, *The Second Creation: Makers of the Revolution in 20th Century Physics*, (New York: Macmillan, 1986), 24.

"It was as if the ground had been pulled out": Einstein, *Autobiographical Notes*, 45.

"According to the assumption": Lightman, *Discoveries*, 55.

"*Lichtquanten*": Overbye, *Einstein in Love*, 120.

"The simplest conception": Lightman, *Discoveries*, 56.

"I insist on the provisional character": Pais, *Subtle Is the Lord*, 383.

"I'm going to give it up": Isaacson, *Einstein*, 122.

"Thank you," Einstein told Besso: Ibid.

"Put your hand on a hot stove": Mayer, *Bite-Sized Einstein*, 65.

"germ of the theory": Isaacson, *Einstein*, 114.

"Picture two observers": Jeremy Bernstein, *Secrets of the Old One*, 71.

Time is "the culprit": Park, *Fire Within the Eye*, 299.

to be "superfluous": Lightman, *Discoveries*, 73.

"It is for the painters": Pais, *Niels Bohr's Times*, 25.

"an extremely sensitive child": Zajonc, *Catching the Light*, 247.

"like a fly in a cathedral": Kumar, *Quantum*, 79.

"[that] we could not proceed at all": Crease and Mann, *Second Creation*, 26.

"stationary state": Pais, *Niels Bohr's Times*, 147.

"an enormous achievement": Ibid., 154.

"gravitational equations": Isaacson, *Einstein*, 201.

"of incomparable beauty": Ibid., 223.

LIGHT ALL ASKEW IN THE HEAVENS: "Light All Askew in the Heavens," *New York Times*, November 10, 1919, 17.

"such a cult has been made": Kumar, *Quantum*, 128.

"For the rest of my life": Sims, *Apollo's Fire*, 29.

"There are therefore now two theories of light": Kumar, *Quantum*, 142.

"pilot waves": Gamow, *Thirty Years That Shook Physics*, 81.

"lifted a corner of the great veil": Isaacson, *Einstein*, 327.

"la *Comédie Française*": Gamow, *Thirty Years That Shook Physics*, 81.

"Professor," a colleague told the physicist Wolfgang Pauli: Kumar, *Quantum*, 164.

"wavicles": Heisenberg, *Physics and Beyond*, 95.

"Physics at the moment": Al-Khalili, *Quantum*, 72.

"what a *Schweinerei*": Crease and Mann, *Second Creation*, 40.

"witches' Sabbath": Kumar, *Quantum*, 66.

"These discontinuities, which we find": Isaacson, *Einstein*, 169.

"It is wrong to think that the task of physics": Pais, *Niels Bohr's Times*, 427.

"God does not play dice": Kumar, *Quantum*, 274.

"It cannot be for us to tell God": Ibid.

"soothing philosophy—or religion?": Pais, *Niels Bohr's Times*, 425.

"This epistemology-soaked orgy": Kumar, *Quantum*, 314.

"We have to live with quantum theory": Ibid., 29.

"But what is light really?": Hawking, *A Stubbornly Persistent Illusion*, 310.

"Spooky action at a distance": Isaacson, *Einstein*, 459.

"All these fifty years of pondering": Knight and Allen, *Concepts of Quantum Optics*, overleaf.

"looking at a strangely beautiful interior": Crease and Mann, *Second Creation*, 48.

CHAPTER 14: CATCHING UP WITH OUR DREAMS

"Tell them to go find a hotter place!": Miller, *Empire of the Stars*, 47.

"People were really at a loss": Ibid., 166.

"The inside of a star": Ibid., 48.

a "random walk": Kopp, "Secrets of the Sun."

"Light brings us news": O'Collins and Meyers, *Light from Light*, 81.

"There is no new thing under the sun": Ecclesiastes 1:9.

"spontaneous emission": Kumar, *Quantum*, 124.

"population inversion": Hecht, *Beam*, 9.

"armed with concentrated sunbeams": Irving, *Historical Tales and Sketches*, 421.

"much as the mirror of a lighthouse": Koch and Taylor "War of the Worlds," 9.

"blasters" and "ray guns": Wilk, *How the Ray Gun Got Its Zap*, 205–6.

"Infrared and Optical Masers": Hecht, *Laser Pioneers*, 13.

"the starting gun": Ibid., 13.

"Conceive a tube": Hecht, *Beam*, 52.

"heat an object": Hecht, *Laser Pioneers*, 124.

"Some rough calculations": Ibid., 114.

"the technological Olympics": Maiman, *Laser Odyssey*, 63.

"relatively high gain coefficient": Ibid., 67.

"We hear that you are still working": Ibid., 102.

"What would Hughes do": Ibid., 64.

"Optical pumping": Hecht, *Beam*, 37.

"A nasty lamp to work with": Maiman, *Laser Odyssey*, 93.

Maiman had his "aha!": Ibid.

"when we got past 950 volts": Ibid., 103.

"The output trace started to shoot up": Ibid.

"numb and emotionally drained": Ibid., 105.

"atomic radio light": Ibid., 114.

"coherent light": Ibid., 55.

"coherent light might become a death ray": Hecht, *Beam*, 192.

LA MAN DISCOVERS SCIENCE FICTION DEATH RAY: Ibid.

"Bullets of Light": Boggs, "Bullets of Light," *New Republic*, March 16, 1963, 5.

"Ruby Ray Guns": "Ruby Ray Guns," *America*, April 6, 1963, 454.

"Light Ray: Fantastic Weapon": "Fantastic Weapon," *U.S. News and World Report*, April 2, 1962, 47.

"Come here, Watson": Hecht, *Beam*, 217.

"beam-directed energy weapons": Raymond, "Air Force Seeking Light Rays to Knock Down Foes' ICBMs," *New York Times*, March 29, 1962, 2.

"You are looking at an industrial laser": Hamilton, "Goldfinger."

"Since the universe began": Poor, "The Laser."

"The laser today stands": Ibid.

"catching up with our dreams": Raymond, "Air Forice Seeks Light Rays to Knock Down Foes' ICBMS," *New York Times*, January 15, 1967, F1.

"Why do you have Feynman Diagrams": Sykes, *No Ordinary Genius*, 85.

"screwy," "dopey," and "absurd": Feynman, "Character of Physical Law."

"There was a moment when I knew": Gleick, *Genius*, 338–39.

"the most brilliant young physicist here": Gleick, *Genius*, photo insert.

"The idea is that you have the quantum theory": Sykes, *No Ordinary Genius*, 70.

"half-assedly thought-out": Gleick, *Genius*, 244.

"the sun breaking through the clouds": Ibid., 394.

"quantum electrodynamics, the strange theory": Feynman, *QED*, 4.

"probability amplitudes": Ibid., 33.

"would be exact to the thickness": Ibid., 7.

"the jewel of physics": Ibid., 8.

"as absurd from the point of view": Ibid., 10.

"make up its mind": Ibid., 19.

"Do not keep asking yourself": Kumar, *Quantum*, 352.

"light pipes": Nardo, *Lasers: Humanity's Magic Light*, 55.

"classical optics": Fox, *Quantum Optics*, 8.

"laser cooling": "Stephen Chu: Laser Cooling and Trapping of Atoms."

EPILOGUE

"the light fantastic": Poor, "The Laser: A Light Fantastic."

"Regardless of the mythological explanation": "Diwali."

"a mesmerizing city of light": "One Destination, Two Holidays: Berlin's Festival of Lights."

"holds great promise": Overbye, "American and 2 Japanese Physicists Share Nobel for Work on LED Lights."

"When we got the kids": Owen, "Dark Side."

"the strange sky": Rong Gong Lin II, "A Desert Plea."

"We finally realized": Ibid.

"a global scale": Cinzano, Falchi, and Elvidge, "First World Atlas of the Artificial Night Sky Brightness," 4.

"Mankind is proceeding to envelop itself": Ibid., 1.

"night's candles": Shakespeare, *Complete Works*, 356.

"glare bombs": Owen, "Dark Side."

"Astronomy Town" and "star parties": Ibid.

"the nightmare spectrum": Betz, "A New Fight for the Night," 49.

"The blue light really has to be suppressed": Ibid.

"essentially a 100 percent rebound": Hanson, "Drowning in Light."

"slow light": "Slow Light; About Light Speed."

"Probably anything you can imagine doing with light": Thomas Miltser, interview with the author, March 31, 2014.

"There's a certain Slant of light": Dickinson, *Complete Poems*, 118.

"a proven seasonal affective disorder treatment": Mayo Clinic, "Tests and Procedures: Light Therapy."

"They just hunkered down": Goldman, "Scientists Discover Anti-anxiety Circuit."

"the most revolutionary thing": Gorman, "Brain Control in a Flash of Light."

"If we can harness this fusion": Frazier, "Fusion Will Be a Huge Energy Breakthrough."

"I thought I understood everything": Wogan, "Controlling Ferro-Magnetic Domains Using Light."

"the Dark Ages": Loeb, "Dark Ages of the Universe," 47.

"Light is something unique, special": Dae Wook Kim, interview with the author, March 31, 2014.

"I was a big fan of the movie": Ibid.

"The Bible is not my physics textbook": Ibid.

"It's taken four hundred years": Ibid.

"I have two sons": Ibid.

"lighting up everything": O'Kelly, *Newgrange: A Concise Guide*, 26.

APPENDIX

"walked along the beam of light": Kelliher, *Experiences Near Death*, 8.

"made of golden light": Corazza, *Near-Death Experiences*, 62.

"floating": Valarino, *On the Other Side of Life*, 58.

"pure, crystal light": Fox, *Spiritual Encounters with Unusual Light Phenomenon*, 47.

"so bright, so radiant": Moody, *Life After Life*, 53.

"not any kind of light": Ibid.

"near-death experience": Ibid., 6.

"a being of light": Ibid., 50.

"white light gliding": Fox, *Spiritual Encounters with Unusual Light Phenomenon*, 39.

"liquid light": Ibid., 35.

"corpse candles": Ibid., 61–62.

"The soul on the point of death": Christopolous, Karakantza, and Levaniouk, *Light and Darkness in Ancient Greek Myth and Religion*, 201.

"Clear Light of Pure Reality": Corazza, *Near-Death Experiences*, 46.

"You learn very quickly": Moody, *Life After Life*, 80.

"divine light": "How to Do the Divine Light Invocation."

"physiologically distinct": Britton and Bootzin, "Near Death Experiences and the Temporal Lobe," 254.

"the light of REM consciousness": Gottlieb, "Back from Heaven."

"Hallucinations cannot provide evidence": Sacks, "Seeing God in the Third Millennium."

"the greatest advances in the rational understanding": Moody, *Life After Life*, 171.

Bibliography and Works Cited

All translations are listed by original author.

"A Conversation with James Turrell," Houston, TX: Contemporary Arts Museum, October 25, 2013. Available on Youtube.

Abdel Haleem, M. A. S., trans., *The Qur'an: Oxford World Classics*, Oxford, UK: Oxford University Press, 2005, Kindle edition.

"About Roden Crater," Roden Crater website (http://rodencrater.com/about; accessed May 12, 2015).

Acocella, Joan, "What the Hell," *New Yorker*, May 27, 2013 (www.newyorker.com /magazine/2013/05/27/what-the-hell; accessed, September 15, 2013).

Aczel, Amir D., *Pendulum: Leon Foucault and the Triumph of Science*, New York: Atria Books, 2003.

Aczel, Amir N., *Descartes's Secret Notebook: A True Tale of Mathematics, Mysticism, and the Quest to Understand the Universe*, New York: Broadway Books, 2005.

Adams, Henry, *Mont St. Michel and Chartres*, New York: Penguin, 1913.

Adamson, Peter, *Al-Kindi*, Oxford, UK: Oxford University Press, 2007.

Adamson, Peter, and Richard D. Taylor, *The Cambridge Companion to Arabic Philosophy*, Cambridge, UK: Cambridge University Press, 2005.

Al-Khalili, Jim, *The House of Wisdom: How Arabic Science Saved Ancient Knowledge and Gave Us the Renaissance*, New York: Penguin, 2011.

———, *Quantum*, London: Weidenfeld & Nicholson, 2003.

American Physical Society, "Michelson and Morley," APS Physics, website (www.aps .org/programs/outreach/history/historicsites/michelson-morley.cfm; accessed May 11, 2015).

Arafat, W., and H. J. J. Winter, "The Light of the Stars—A Short Discourse by Ibn al-Haytham," *British Journal for the History of Science* 5, no. 3 (1971): 282–88.

Arago, François, *Biographies of Distinguished Scientific Men*, Boston: Ticknor & Fields, 1859. Available at Google Books.

Armstrong, Karen, *A History of God: The 4,000 Year Quest of Judaism, Christianity and Islam*, New York: Ballantine Books, 1993.

———, *The Case for God*, New York: Alfred A. Knopf, 2009.

Assumen, Jes P., *Manichaean Literature: Representative Texts Chiefly from Middle Persian and Parthian Writings*, New York: Scholars' Facsimiles & Reprints, 1975.

Attlee, James, *Nocturne: A Journey in Search of Moonlight*, Chicago: University of Chicago Press, 2011.

Aquinas, Thomas, *Summa Theologica*, trans., Fathers of the English Dominican Province, Grand Rapids, MI: Christian Classics Ethereal Library, 2009, Kindle edition.

Ball, Philip, *Universe of Stone: A Biography of Chartres Cathedral*, New York: HarperCollins, 2009.

Bambrough, Renford, ed., *The Philosophy of Aristotle*, New York: Mentor Books, 1963.

Barasch, Moshe, *Light and Color in the Italian Renaissance Theory of Art*, New York: New York University Press, 1978.

Barber, X. Theodore, "Phantasmagorical Wonders: The Magic Lantern Ghost Show in Nineteenth Century America," *Film History* 3, no. 2 (1989): 73–86.

Barnes, Jonathan, ed., *Early Greek Philosophy*, New York: Penguin, 1987.

Barocas, V., *The Nature of Light: An Historical Survey*, trans. Vasco Ronchi, London: Heinemann, 1970.

Beare, J. I., trans., *The Complete Aristotle*, Adelaide, Australia: Feedbooks.com, 2011, Kindle edition.

Behrens-Abouseif, Doris, *The Minarets of Cairo: Islamic Architecture from the Arab Conquest to the end of the Ottoman Period*, Cairo: American University in Cairo Press, 1985.

Bernstein, Jeremy, *Secrets of the Old One: Einstein 1905*, New York: Copernicus, 2006.

Binyon, Laurence, trans., *The Portable Dante*, New York: Penguin Books, 1947.

Birch, Dinah, ed., *Ruskin on Turner*, Boston, Toronto, London: Little, Brown, 1990.

Blake, William, *Complete Works of William Blake*, Delphi Poets, 2012, Kindle edition.

Bodanis, David, *$E=mc^2$: A Biography of the World's Most Famous Equation*, New York: Berkley Books, 2000.

Boggs, W. E., "Bullets of Light," *New Republic*, March 16, 1963, 5.

Bondanella, Julia Conaway, and Peter Bondanella, trans., *Lives of the Artists*, by Giorgio Vasari, Oxford, UK: Oxford University Press, 1991, Kindle edition.

Britton, Willoughby B. and Richard R. Bootzin, "Near Death Experiences and the Temporal Lobe," *Psychological Science* 15, no. 4 (April 2004); 254–58.

Brox, Jane, *Brilliant: The Evolution of Artificial Light*, Boston and New York: Houghton Mifflin Harcourt, 2010.

Buchwald, Jed Z., *The Rise of the Wave Theory of Light: Optical Theory and Experiment in the Early Nineteenth Century*, Chicago and London: University of Chicago Press, 1989.

Burke, Robert Belle, trans., *The Opus Majus of Roger Bacon, Volume II*, Philadelphia: University of Pennsylvania Press, 1928.

Burnett, D. Graham, *Descartes and the Hyperbolic Quest: Lens Making Machines and Their Significance in the Seventeenth Century*, Philadelphia: American Philosophical Society, 2005.

Buttrick, George, *The Interpreter's Bible*, Nashville: Abingdon Press, 1954.

Byron, Gordon George (Lord Bryon), *Don Juan*, Seattle: Amazon Digital Services, 2012, Kindle edition.

Cahn, Steven M., ed., *Classics of Western Philosophy*, Indianapolis: Hackett Publishing Company, 1977.

Callen, Anthea, *The Art of Impressionism: Painting Technique and the Making of Modernity*, New Haven and London: Yale University Press, 2000.

———, *Techniques of the Impressionists*, London: QED Publishing, 1982.

Campbell, Joseph, and Bill Moyers, *The Power of Myth*, New York: Doubleday, 1991.

Campbell, Louis, *The Life of James Clerk Maxwell*, Ann Arbor, MI: University of Michigan Press, 1882.

Capra, Fritjof, *Leonardo: Inside the Mind of the Great Genius of the Renaissance*, New York: Doubleday, 2007.

———, *The Tao of Physics: An Exploration of the Parallels Between Modern Physics and Eastern Mysticism*, Boston: Shambhala Publications, 2010.

Cary, Henry F., trans., *Paradiso*, Internet Sacred Text Archive (www.sacred-texts.com/chr/dante/pa23.htm; accessed September 15, 2013).

Cashford, Jules, *The Moon: Myth and Image*, New York: Four Walls Eight Windows, 2003.

Chadwick, Henry, trans., *Augustine: Confessions*, Oxford, UK: Oxford World Classics, 1998, Kindle edition.

Chen, Cheng-Yih, ed., *Science and Technology in Chinese Civilization*, Singapore: World Scientific, 1987.

Cheney, Ian, director, *The City Dark*, New York: Edgeworx Studios, 2011.

Christianson, Gale E., *In the Presence of the Creator: Isaac Newton and His Times*, New York: Macmillan, 1984.

Christopolous, Menelaos, Efimia D. Karakantza, and Olga Levaniouk, *Light and Darkness in Ancient Greek Myth and Religion*, New York, Toronto, and Plymouth, UK: Rowman & Littlefield, 2010.

Cinzano, P., T. Falchi, and C. D. Elvidge, "The First World Atlas of the Artificial Night Sky Brightness," *Monthly Notices of the Royal Astronomical Society*, August 3, 2001, 1–24.

Clark, Robin, ed., *Phenomenal: California Light, Space, Surface*, Berkeley and Los Angeles: University of California Press, 2011.

Clarke, Desmond M., trans., *René Descartes: Discourse on the Method and Related Writings*, New York: Penguin Classics, 1999.

Clegg, Brian, *The First Scientist: A Life of Roger Bacon*, New York: Carroll & Graf, 2003.

Cohen, Richard, *Chasing the Sun: The Epic Story of the Star that Gives us Life*, New York: Random House, 2010.

Collins, Theresa M., and Lisa Gitelman, *Thomas Edison and Modern America: A Brief History with Documents*, Boston and New York: Bedford/St. Martin's, 2002.

Corazza, Ornella, *Near-Death Experiences: Exploring the Mind-Body Connection*, London and New York: Routledge, 2008.

Crease, Robert P., and Charles C. Mann, *The Second Creation: Makers of the Revolution in 20th Century Physics*, New York: Macmillan, 1986.

Crew, Henry, ed., *The Wave Theory of Light: Memoirs by Huygens, Young, and Fresnel*, New York: American Book Company, 1900.

Crompton, Samuel Willard, and Michael J. Rhein, *The Ultimate Book of Lighthouses*, Rowayton, CT: Thunder Bay Press, 2000.

Crosby, Sumner McKnight, Jane Hayward, Charles T. Little, and William D. Wixom, *The Royal Abbey of Saint-Denis in the Time of Abbot Suger (1122–1151)*, New York: Metropolitan Museum of Art, 1981.

Cummings, E. E., *Complete Poems: 1913–1962*, New York and London: Harcourt Brace Jovanovich, 1963.

Daguerre, Louis Jacques-Mandé, *History and Practice of Photogenic Drawing on the True Principles of the Daguerreotype*, trans. J. S. Memes, 3rd edition, London: Smith, Elder and Co., and Edinburgh: Adam Black and Co., 1839. Available at Google Books.

DeKroon, Pieter-Rim, director, *Dutch Light*, DVD, Dutch Light Films, Amsterdam, 2003.

Della Porta, Giambattista, *Natural Magick: A Neapolitane in Twenty Books*, London: Thomas Young and Samuel Speed, 1658.

Dickinson, Emily, *The Complete Poems of Emily Dickinson*, Thomas H. Johnson, ed., Boston, Toronto: Little, Brown, 1960.

"Diwali," Religions, BBC, website (www.bbc.co.uk/religion/religions/hinduism/holydays/diwali.shtml; accessed May 11, 2015).

Dolnick, Edward, *The Clockwork Universe: Isaac Newton, the Royal Society, and the Birth of the Modern World*, New York: HarperPerennial, 2011.

Dundes, Alan, ed., *Sacred Narrative: Readings in the Theory of Myth*, Berkeley: University of California Press, 1984.

Dustan, Bernard, *Painting Methods of the Impressionists*, New York: Watson-Guptill, 1976.

Easwaran, Eknath, trans., *The Bhagavad Gita*, 1954; reprint, Berkeley: Nilgiri Press, 2007, Kindle edition.

Eck, Diana L., *Banaras, City of Light*, New York: Columbia University Press, 1982.

"Edison's Electric Light," *New York Times*, Sept. 5, 1882, 8.

Einstein, Albert, *Autobiographical Notes*, Chicago: Open Court, 1979.

"Einstein and Lasers," Advances in Atomic Physics, website (www.psc.edu/science/Eberly/Eberly.html; accessed May 11 2015).

Eiseley, Loren, *The Immense Journey: An Imaginative Naturalist Explores the Mysteries of Man and Nature*, New York: Vintage, 1959.

Ekrich, A. Roger, *At Day's Close: Night in Times Past*, New York: W.W. Norton, 2003.

"Electric Illumination, *New York Times*, June 15, 1879, 5.

Eliade, Mircea, "Spirit, Light, and Seed," *History of Religions* 11, no. 1 (1971): 1–30.

———, *A History of Religious Ideas, Vol. 3: From Muhammad to the Age of Reforms*, trans. Alf Hiltebeitel and Diane Apostolos-Cappadona, Chicago and London: University of Chicago Press, 1985.

Eliot, T. S., *The Complete Poems and Plays: 1909–1950*, New York: Harcourt, Brace & World, 1971.

Evans, Robert, "Blast from the Past," *Smithsonian*, July 2002.

Faraday, Michael, *The Chemical History of a Candle*, London: Chatto & Windus, 1908, Kindle edition.

Faraday, Michael, "Experimental Researches in Electricity, Nineteenth Series," *Philosophical Transactions of the Royal Society of London* 136 (1846): 2.

———, *Romanticism: A Very Short Introduction*, Oxford, UK: Oxford University Press, 2010.

Ferris, Timothy, *Coming of Age in the Milky Way*, New York: Doubleday, 1988.

Feynman, Richard, "The Character of Physical Law," The Messenger Lectures, Cornell University, Ithaca, NY, 1964. Available on YouTube.

———, *QED: The Strange Theory of Light and Matter*, Princeton: Princeton University Press, 1985.

Filipczak, Z. Zaremba, "New Light on Mona Lisa: Leonardo's Optical Knowledge and His Choice of Lighting," *Art Bulletin* 59, no. 4 (Dec., 1977): 518–23.

Forbes, Nancy, and Basil Mahon, *Faraday, Maxwell and the Electromagnetic Field: How Two Men Revolutionized Physics*, Amherst, NY: Prometheus Books, 2014, Kindle edition.

Fox, Douglas A., "Darkness and Light: The Zoroastrian View," *Journal of the American Academy of Religion* 35, no. 2 (June 1967): 129–37.

Fox, Mark, *Quantum Optics: An Introduction*, Oxford UK: Oxford University Press, 2006.

———, *Spiritual Encounters with Unusual Light Phenomena: Lightforms*, Cardiff, Wales: University of Wales Press, 2008.

Frazier, Tim, "Fusion Will Be a Huge Energy Breakthrough, Says National Ignition Facility CIO," *Forbes* online, September 29, 2014 (www.forbes.com/sites/netapp/2014/09/29/fusion-clean-energy-nif-cio/; accessed October 27, 2014).

Freeberg, Ernest, *The Age of Edison: Electric Light and the Invention of Modern America*, New York: Penguin, 2013.

Frova, Andrea, Mariapiera Marenzana, and Jim McManus, trans., *Thus Spoke Galileo: The Great Scientist's Ideas and Their Relevance to the Present Day*, Oxford, UK: Oxford University Press, 2006.

Gamow, George, *Thirty Years That Shook Physics: The Story of Quantum Theory*, Mineola, NY: Dover Publications, 1966.

Gardner, Martin, *Relativity Simply Explained*, Mineola, NY: Dover Publications, 1972.

Gaukroger, Stephen, *Descartes: An Intellectual Biography*, Oxford, UK: Clarendon Press, 1995.

Geiringer, Karl, *Haydn: A Creative Life in Music*, New York: W. W. Norton, 1946.

Gernsheim, Helmut, and Alison Gernsheim, *L. J. M. Daguerre: The History of the Diorama and the Daguerreotype*, London: Secker & Warburg, 1956.

Gleick, James, *Genius: The Life and Science of Richard Feynman*, Princeton: Princeton University Press, 1992.

———, *Isaac Newton*, New York: Random House, 2004.

Goldman, Bruce, "Scientists Discover Anti-Anxiety Circuit in Brain Region Considered the Seat of Fear," Stanford Medicine News Center, March 9, 2011 (http://med.stanford.edu/news/all-news/2011/03/scientists-discover-anti-anxiety-circuit-in-brain-region-considered-the-seat-of-fear.html; accessed Oct. 2, 2014).

Goldman, Martin, *The Demon in the Ether: The Story of James Clerk Maxwell*, Edinburgh: Paul Harris Publishing, 1983.

Gorman, James, "Brain Control in a Flash of Light," *New York Times*, April 14, 2014.

Gottlieb, Robert, "Back from Heaven—The Science," *New York Review of Books*, November 6, 2014 (www.nybooks.com/articles/archives/2014/nov/06/back-heaven-science/?page=1; accessed October 22, 2014).

Govan, Michael, and Christine Y. Kim, *James Turrell: A Retrospective*, Los Angeles and London: Prestel Publishing, 2013.

Grafton, Anthony, *Leon Battista Alberti: Master Builder of the Italian Renaissance*, New York: Hill & Wang, 2000.

Graham, A. C., and Nathan Sivin, "A Systematic Approach to Mohist Optics, ca. 300 BCE," in Shigeru Nakayama, and Nathan Sivin, eds., *Chinese Science*, Cambridge, MA, and London: MIT Press, 1973.

Grayling, A. C., *Descartes: The Life and Times of a Genius*, New York: Walker & Co., 2005.

Grayson, Cecil, ed. and trans., *Leon Battista Alberti: On Painting and Sculpture*, New York: Phaidon, 1972.

Griffith, Ralph T. H., trans., *The Complete Rig Veda*, Seattle: Classic Century Works, 2012, Kindle edition.

Griffith, Tom, trans., *Plato, The Republic*, Cambridge, UK: Cambridge University Press, 2000.

Hall, A. Rupert, *All Was Light: An Introduction to Newton's Opticks*, Oxford, UK: Clarendon Press, 1993.

———, "Sir Isaac Newton's Notebook, 1661–1665" *Cambridge Historical Journal* 9, no. 2 (1948): 239–50.

Hamilton, Guy, director, *Goldfinger*, 20th Century Fox, 1964.

Hamilton, James, *A Life of Discovery: Michael Faraday, Giant of the Scientific Revolution*, New York: Random House, 2002.

Hamilton, James, *Turner: A Biography*, New York: Random House, 2003.

Hanna, John, trans., *The Elements of Sir Isaac Newton's Philosophy, by Mr. Voltaire*, London: Medicine, Science, and Technology, 1738.

Hanson, Dirk, "Drowning in Light," *Nautilus*, March 2014 (http://m.nautil.us/issue/11/light; accessed November 1, 2014).

Hastie, W., trans., *Kant's Cosmogony*, Glasgow: James Maclehose & Sons, 1900, Kindle edition.

Hawking, Stephen, ed., *A Stubbornly Persistent Illusion: The Essential Scientific Works of Albert Einstein*, Philadelphia, London: Running Press, 2007.

Haydn, Joseph, *Die Schöpfung: Ein Oratorium in Musik*, Munich: G. Henle, 2009.

Haydocke, Richard, trans., *A Tracte Containing the Arte of Curious Paintings Caruinge Buildinge Written First In Italian by Io: Paul Laumatius, Painter of Milan*, Oxford, 1598.

Headlam, Cecil, *The Story of Chartres*, London: J. M. Dent, 1930.

Hecht, Jeff, *Beam: The Race to Make the Laser*, Oxford, UK: Oxford University Press, 2005.

———, *Laser Pioneers*: rev. ed, Boston: Harcourt, Brace, Jovanovich, 1992.

Heisenberg, Werner, *Physics and Beyond: Encounters and Conversations*, New York: Harper Torchbooks, 1972.

Hibbard, Howard, *Caravaggio*, New York: Harper & Row, 1983.

Hill, Edmund, trans., *Augustine: On Genesis*, Hyde Park, NY: New City Press, 2002.

Hillenbrand, Robert, *Islamic Architecture: Form, Function, and Meaning*, New York: Columbia University Press, 2004.

Hills, Paul, *The Light of Early Italian Painting*, New Haven and London: Yale University Press, 1987.

Hirsh, Diana, *The World of Turner—1775–1851*, New York: Time-Life Books, 1969.

Hockney, David, *Secret Knowledge: Rediscovering the Lost Techniques of the Old Masters*, New and expanded edition, New York: Penguin, 2006.

Hollander, Robert and Jean, trans., *Dante: Paradiso*, New York: Anchor Books, 2007.

Holmes, Richard, *The Age of Wonder: How the Romantic Generation Discovered the Beauty and Terror of Science*, New York: Random House, 2008.

———, *Shelley: The Pursuit*, New York: NYRB Classics, 1994.

"How to Do the Divine Light Invocation," Wicca Spirituality, website (www.wicca-spirituality.com/the_divine_light_invocation.html, accessed May 21, 2015).

Hugo, Victor, *Les Misérables*, trans. Norman Denny, New York: Penguin, 1980.

———, *The Hunchback of Notre-Dame*, trans. Walter J. Cobb, New York: New American Library, 1964.

Hume, Robert Ernest, ed. and trans., *The Thirteen Principal Upanishads*, London: Oxford University Press, 1971.

Huygens, Christiaan, *Treatise on Light*, trans. Silvanus P. Thompson, Chicago: University of Chicago Press, 2011 Kindle edition.

Hyman, Arthur, and James J. Walsh, eds., *Philosophy in the Middle Ages: The Christian, Islamic, and Jewish Traditions*, New York, Evanston, and London: Harper & Row, 1967.

Irani, D. J., "The Gathas: The Hymns of Zarathushtra" (www.zarathushtra.com/z/gatha/dji/The%20Gathas%20-%20DJI.pdf; accessed, May 21, 2015).

Irving, Washington, *Historical Tales and Sketches*, New York: Library of America, 1983.

Isaacson, Walter, *Einstein: His Life and Universe*, New York, London, Toronto, Sydney: Simon & Schuster, 2007.

Jager, Colin, *The Book of God: Secularization and Design in the Romantic Era*, Philadelphia: University of Pennsylvania Press, 2007.

James, William, *The Varieties of Religious Experience*, New York: Modern Library, 1936.

Jehl, Frances, *Menlo Park Reminiscences*, Dearborn, MI: Edison Institute, 1936.

Jones, Colin, *Paris: The Biography of a City*, New York: Viking, 2005.

Jonnes, Jill, *Empires of Light: Edison, Tesla, Westinghouse, and the Race to Electrify the World*, New York: Random House, 2003.

Jowett, Benjamin, trans., *Plato: The Complete Works*, Kirkland, WA: Latus ePublishing, 2011, Kindle edition.

———, *The Portable Plato*, New York: Penguin Books, 1981.

Kaku, Michio, *Physics of the Impossible: A Scientific Exploration into the World of Phasers, Force Fields, Teleportation, and Time Travel*, New York: Random House, 2008.

Kapstein, Matthew T., ed., *The Presence of Light: Divine Radiance and Religious Experience*, Chicago and London: University of Chicago Press, 2004, 286–87.

Keats, John, *John Keats: Complete Works*, Delphi Classics, 2012, Kindle edition.

Kelliher, Allan, *Experiences Near Death: Beyond Medicine and Religion*, New York and Oxford, UK: Oxford University Press, 1996.

Kemp, Martin, *The Science of Art: Optical Themes in Western Art from Brunelleschi to Seurat*, New Haven and London: Yale University Press, 1990.

Kennedy, Hugh, *When Baghdad Ruled the Muslim World: The Rise and Fall of Islam's Great Dynasty*, Cambridge, MA: Da Capo Press, 2005.

Kepler, Johannes, *Optics: Paralipomena to Witelo and the Optical Part of Astronomy*, trans. William H. Donohue, Santa Fe, NM: Green Lion Press, 2000.

Kheirandish, Elaneh, "The Many Aspects of 'Appearances': Arabic Optics to 950 AD," in Jan P. Hogendijk and Abdelhamid I. Sabra, eds., *The Enterprise of Science in Islam: New Perspectives*, Cambridge, MA: MIT Press, 2003.

Kheirandish, Elaneh, trans. and ed., *The Arabic Edition of Euclid's "Optics,"* New York, Springer-Verlag, 1999.

"King of Exalted, Glorious Sutras Called the Exalted, Sublime Golden Light, The" Foundation for the Preservation of Mahayana Tradition (www.fpmt.org/education/teachings/sutras/golden-light-sutra.html; accessed May 15, 2013).

Kirk, G. S., J. E. Raven, and M. Schofield, *The Pre-Socratic Philosophers*, 2nd ed., Cambridge, UK: Cambridge University Press, 1983.

Kline, A. S., trans., "Dante: The Divine Comedy; Paradiso, Cantos XXII–XXVIII," Poetry In Translation, website (www.poetryintranslation.com/PITBR/Italian/DantPar22to28.htm#_Toc64099971; accessed May 10, 2015).

Knight, P. L., and L. Allen, *Concepts of Quantum Optics*, London: Pergamon Press, 1983.

Koch, Howard, and Anne Froelick Taylor, "The War of the Worlds," radio drama, *Mercury Theater of the Air*, CBS, first broadcast October 30, 1938.

Koestler, Arthur, *The Sleepwalkers: A History of Man's Changing View of the Universe*, New York: Grosset & Dunlap, 1963.

Kopp, Duncan, director, "Secrets of the Sun," *Nova*, PBS, first broadcast April 25, 2012.

Koslofsky, Craig, *Evening's Empire: A History of the Night in Early Modern Europe*, Cambridge, UK: Cambridge University Press, 2011.

Kriwaczek, Paul, *In Search of Zarathustra: The First Prophet and the Ideas That Changed the World*, New York: Alfred A. Knopf, 2003.

Kübler-Ross, Elisabeth, *On Life After Death*, Berkeley: Celestial Arts, 1991.

Kumar, Manjit, *Quantum: Einstein, Bohr, and the Great Debate About the Nature of Reality*, New York, London: W. W. Norton, 2008.

"Landmarks: A Conversation with James Turrell," University of Texas, Austin, TX, October 18, 2013 (www.youtube.com/watch?v=nsGxFiFsxY8; accessed May 12, 2015).

Landon, H. C. Robert, *Haydn: Chronicle and Works*, London: Thames & Hudson, 1994.

"The Larger Sutra of Immeasurable Life, Part 1," Pure Land Buddhist Scriptures (http://buddhistfaith.tripod.com/purelandscriptures/id2.html, accessed May 14, 2013).

Le Gall, Guillaume, *La Peinture Mecanique—Le Diorama de Daguerre*, Paris: Mare & Martin, 2013.

Leeming, David, and Margaret Leeming, *A Dictionary of Creation Myths*, New York and Oxford UK: Oxford University Press, 1994.

Levitt, Theresa, "Editing Out Caloric: Fresnel, Arago, and the Meaning of Light," *British Journal for the History of Science* 33, no. 1 (March 2000): 49–65.

———, *A Short Bright Flash: Augustin Fresnel and the Birth of the Modern Lighthouse*, New York, London: W. W. Norton, 2013.

Lewis, David, trans., *The Life of St. Teresa of Jesus*, London, New York: Thomas Baker Benziger Bros., 1904, Kindle edition.

Liedtke, Walter, *A View of Delft: Vermeer and His Contemporaries*, Zwolle: Waanders Printers, 2000.

"Light All Askew in the Heavens," *New York Times*, Nov. 10, 1919, 17.

"Light Ray: Fantastic Weapon," *U.S. News and World Report*, April 2, 1962, 47.

Lightman, Alan, *The Discoveries: Great Breakthroughs in 20th Century Science, Including the Original Papers*, New York: Random House, 2005.

Lindberg, David C., *Theories of Vision: From Al-Kindi to Kepler*, Chicago and London: University of Chicago Press, 1976.

Livingston, Dorothy Michelson, *The Master of Light: A Biography of Albert A. Michelson*, New York: Charles Scribner's Sons, 1973.

Loeb, Abraham, "The Dark Ages of the Universe," *Scientific American*, November 2006.

Long, Charles H., *ALPHA: The Myths of Creation*, New York: George Braziller, 1963.

Loudon, Rodney, *The Quantum Theory of Light*, Oxford, UK: Oxford University Press, 1973.

Lowry, Bates, and Isabel Barrett, *The Silver Canvas: Daguerreotype Masterpieces from the J. Paul Getty Museum*, Los Angeles: J. Paul Getty Museum, 1998.

Mahoney, Michael Sean, trans., *Le Monde, ou Traité de la lumieré,* New York: Abaris Books, 1979.

Maiman, Theodore H., *The Laser Odyssey*, Blaine, WA: Laser Press, 2000.

Mandelbaum, Allen, trans., *Dante: Inferno*, New York: Quality Paperback Book Club, 1984.

———, *Dante: Paradiso*, New York: Quality Paperback Book Club, 1984.

———, *Dante: Purgatorio*, New York: Quality Paperback Book Club, 1984.

Matthaei, Ruppert, trans., *Goethe's Color Theory*, New York: Van Nostrand Reinhold Co., 1970.

Maxim, Hiram Percy, *A Genius in the Family*, New York: Dover Publications, 1962.

Maxwell, James Clerk, "A Dynamical Theory of the Electromagnetic Field," *Philosophical Transactions of the Royal Society of London* 155 (1865): 459–512.

Maxwell, James Clerk, *Five of Maxwell's Papers*, Seattle, WA: Amazon Digital Services, Kindle edition.

Maxwell, James Clerk, "To the Chief Musician upon Nabla: A Tyndallic Ode" (www.poetryfoundation.org/poem/175048; accessed May 14, 2014).

Mayer, Jerry, ed. *Bite-Sized Einstein: Quotations on Just About Everything from the Greatest Mind of the 20th Century*, New York: Macmillan, 1996.

Mayo Clinic, "Tests and Procedures: Light Therapy," Mayo Clinic website (www.mayoclinic.org/tests-procedures/light-therapy/basics/definition/prc-20009617; accessed May 12, 2015).

McClure, James Baird, *Edison and His Inventions: Including the Many Incidents, Anecdotes, and Interesting Particulars Connected with the Early and Late Life of the Great Inventor*, Chicago: Rhodes & McClure Publishing, 1889. Available at Google Books.

McDannell, Colleen, and Bernard Lang, *Heaven: A History*, New Haven: Yale University Press, 2001.

McEvoy, James, "The Metaphysics of Light in the Middle Ages," *Philosophical Studies* 26 (1979): 126–45.

McEvoy, James, *Robert Grosseteste*, Oxford, UK: Oxford University Press, 2000.

McEvoy, J. P., and Oscar Zarate, *Introducing Quantum Theory*, Cambridge, UK: Icon Books, 1999.

McLuhan, Marshall, *Understanding Media: The Extensions of Man*, New York: McGraw-Hill, 1964.

McMahon, A. Philip, trans., *Treatise on Painting*, by Leonardo da Vinci, Princeton: Princeton University Press, 1956.

Michelson, Albert, "Experimental Determination of the Velocity of Light" (www.gutenberg.org/files/11753/11753-0.txt; accessed June 23, 2014).

———, *Light Waves and Their Uses*, Chicago: University of Chicago Press, 1903. Available at Google Books.

Millard, Andre, *Edison and the Business of Invention*, Baltimore and London: Johns Hopkins University Press, 1990.

Miller, Arthur I., *Empire of the Stars: Obsession, Friendship, and Betrayal in the Quest for Black Holes*, Boston: Houghton Mifflin, 2005.

Milton, John, *Paradise Lost* ("The Project Gutenberg Ebook of Paradise Lost, by John Milton," www.gutenberg.org/files/26/26.txt; accessed February 4, 2013).

Moffitt, John F., *Caravaggio in Context: Learned Naturalism and Renaissance Humanism*, Jefferson, NC, and London: McFarland & Co., 2004.

Mollon, J. D., "The Origin of the Concept of Interference," *Philosophical Transactions: Mathematical, Physical, and Engineering Sciences* 360, no. 1794: 807–19.

Moody, Raymond A., Jr., *Life After Life: The Investigation of a Phenomenon—Survival After Bodily Death*, 25th anniversary edition, New York: HarperCollins, 2001.

Morley, Henry, trans., *Letters on England by Voltaire*, Coventry: Cassel & Co., 1894, Kindle edition.

"Mythbusters, Presidential Challenge," *Mythbusters*, season 9, episode 10, December 8, 2010.

Nagar, Shanti Lai, *Indian Gods and Goddesses: The Early Deities from Chalcolithic to Beginning of Historical Period*, Delhi, India: BR Publishing, 2004.

Nardo, Don, *Lasers: Humanity's Magic Light*, San Diego: Lucent Books, 1990.

Newton, Isaac, *General Scholium* (http://isaac-newton.org/general-scholium/; accessed June 3, 2014).

Newton, Isaac, *Opticks: Or A Treatise of the Reflections, Refractions, Inflections, and Colours of Light*, 4th ed., London: William Innys, 1730, Kindle edition.

Nicholson, Marjorie Hope, *Newton Demands the Muse: Newton's Opticks and the Eighteenth Century Poets*, Princeton: Princeton University Press, 1946.

Nylan, Michael, "Beliefs About Seeing: Optics and Moral Technologies in Early China," *Asia Major* 21, no. 1: 89–132.

O'Collins, Gerald S. J., and Mary Ann Meyers, *Light from Light: Scientists and Theologians in Dialogue*, Grand Rapids, MI, and Cambridge, UK: Wm. B. Erdmans, 2012.

"One Destination, Two Holidays: Berlin's Festival of Lights," *Sunday (London) Times*, September 7, 2014 (www.thesundaytimes.co.uk/sto/travel/Holidays/article1454796.ece; accessed October 7, 2014).

Orlet, Christopher, "Famous Last Words," *Vocabula Review*, online magazine, July–August 2002 (www.vocabula.com/index.asp; accessed February 24, 2014).

Overbye, Dennis, "American and 2 Japanese Physicists Share Nobel for Work on LED Lights," *New York Times*, October 7, 2014.

———, *Einstein in Love: A Scientific Romance*, New York: Viking, 2000.

Owen, David, "The Dark Side," *New Yorker*, August 20, 2007 (www.newyorker.com/magazine/2007/08/20/the-dark-side-2; accessed October 8, 2014).

O'Kelly, Claire, *Newgrange: A Concise Guide*, Dublin: Eden Publications, 2013.

Pais, Abraham, *Niels Bohr's Times: In Physics, Philosophy, and Polity*, Oxford, UK: Clarendon Press, 1991.

———, *Subtle Is the Lord: The Science and Life of Albert Einstein*, Oxford, UK: Oxford University Press, 1982.

Panofsky, Erwin, *Abbot Suger on the Abbey Church of St. Denis and Its Art Treasures*, Princeton: Princeton University Press, 1976.

Parisinou, Eva, *The Light of the Gods: The Role of Light in Archaic and Classical Greek Culture*, London: Gerald Duckworth, 2000.

Park, David, *The Fire Within the Eye: A Historical Essay on the Nature and Meaning of Light*, Princeton: Princeton University Press, 1998.

Parronchi, Alessandro, *Caravaggio*, Rome: Edizioni Medusa, 2002.

Pelikan, Jaroslav, *The Light of the World: A Basic Image in Early Christian Thought*, New York: Harper & Bros., Publishers, 1962.

Petroski, Henry, *Success Through Failure: The Paradox of Design*, Princeton: Princeton University Press, 2006.

Pissarro, Joachim, *Monet's Cathedral: Rouen 1892–1894*, New York: Alfred A. Knopf, 1990.

Poor, Peter, director, "The Laser: A Light Fantastic," narrated by Walter Cronkite, *Twentieth Century*, CBS News, 1967 (http://yttm.tv/v/8810; accessed May 12, 2015).

Pope, Alexander, *Delphi Complete Works of Alexander Pope*, Delphi Poets Series, 2012, Kindle edition.

———, trans., *The Iliad*, Seattle: Amazon Digital Sources, 2010, Kindle edition.

Pseudo-Dionysius the Areopagite, *De Coelesti Hierarchia*, London: Limovia.net, 2012, Kindle edition.

Rabinowitch, Eugene, "An Unfolding Discovery," *Proceedings of the National Academy of Sciences of the United States of America* 68, no. 11 (November 1971), 2875–76.

Randall, Lisa, *Warped Passages: Unraveling the Mysteries of the Universe's Hidden Dimensions*, New York: HarperCollins, 2005.

Ray, Praphulla Chandra, *A History of Hindu Chemistry: From the Earliest Times to the Middle of the Sixteenth Century CE, Vol. 1.*, Calcutta: Bengal Chemical and Pharmaceutical Works, Ltd., 1903.

Razavi, Mehdi Amin, *Suhrawardi and the School of Illumination*, Surrey, UK: Curzon Press, 1997.

Reeves, Eileen, *Galileo's Glassworks: The Telescope and the Mirror*, Cambridge, MA: Harvard University Press, 2008.

Reston, James, Jr., *Galileo: A Life*, New York: HarperCollins, 1994.

Richter, Irma A., trans., *Leonardo da Vinci: Notebooks*, Oxford, UK: Oxford University Press, 2008.

Riedl, Clare C., trans., *Robert Grosseteste on Light*, Milwaukee: Marquette University Press, 1978.

Robb, Peter, *M—The Man Who Became Caravaggio*, New York: Henry Holt, 1998.

Robinson, Andrew, *The Last Man Who Knew Everything: Thomas Young, the Anonymous Genius Who Proved Newton Wrong and Deciphered the Rosetta Stone, Among Other Surprising Feats*, New York: Penguin, 2007.

Rodin, Auguste, "Gothic in the Cathedrals and Churches of France," *North American Review* 180, no. 579 February 1905): 219–29.

Rodis-Lewis, Genevieve, *Descartes: His Life and Thought*, trans. Jane Marie Todd, Ithaca, NY: Cornell University Press, 1998.

Rong-Gong Lin II, "A Desert Plea: Let There Be Darkness," *LA Times*, January 4, 2011.

Roque, Georges, "Chevreul and Impressionism: A Reappraisal," *Art Bulletin* 7, no. 1 (1996): College Art Association, 26–39.

Rose, Charlie, "Interview with James Turrell," July 1, 2013 (www.youtube.com/watch?v=_bvg6kaWIeo and www.youtube.com/watch?v=1-gmHA7KbcU; accessed May 12, 2015).

Rubenstein, Richard E., *When Jesus Became God: The Struggle to Define Christianity During the Last Days of Rome*, San Diego, New York, London: Harcourt, 1999.

"Ruby Ray Guns," *America*, April 6, 1963, 454.

Rudnick, "The Photogram—A History," Photograms, Art and Design, website (www.photograms.org/chapter01.html; accessed December 11, 2014).

Rudolph, Conrad, *Artistic Change at Saint-Denis: Abbot Suger's Program and the Early Twelfth-Century Controversy over Art*, Princeton: Princeton University Press, 1990.

Rudolph, Kurt, *Gnosis: The Nature and History of Gnosticism*, San Francisco: Harper & Row, 1977.

Sabra, Abdelhamid I., trans., "Ibn al-Haytham's Revolutionary Project in Optics," in Jan P. Hogendijk and Abdelhamid I. Sabra, eds. *The Enterprise of Science in Islam: New Perspectives*, Cambridge, MA, and London: MIT Press, 2003.

———, *Optics, Astronomy, and Logic*, Brookfield, VT: Ashgate Publishing, 1994.

———, *The Optics of Ibn Al-Haytham, Books I-III, On Direct Vision*, London: Warburg Institute, 1989.

Sacks, Oliver, "Seeing God in the Third Millennium," *The Atlantic*, December 12, 2012 (www.theatlantic.com/health/archive/2012/12/seeing-god-in-the-third-millennium/266134/; October 6, 2014).

Schama, Simon, *Citizens: A Chronicle of the French Revolution*, New York: Alfred A. Knopf, 1989.

———, *Rembrandt's Eyes*, New York: Alfred A. Knopf, 1999.

Schneider, Pierre, *The World of Manet: 1832–1883*, New York: Time-Life Books, 1968.

Scott, Robert A., *The Gothic Enterprise: A Guide to Understanding the Medieval Cathedral*, Berkeley, Los Angeles, London: University of California Press, 2003.

Seamon, David, and Arthur Zajonc, eds., *Goethe's Way of Science: A Phenomenology of Nature*, Albany, NY: State University of New York Press, 1998.

Sepper, Dennis L., *Goethe Contra Newton: Polemics and the Project for a New Science of Color*, Cambridge, UK: Cambridge University Press, 1988.

Shakespeare, William, *The Complete Works*, compact edition, Oxford, UK: Oxford University Press, 1988.

Shanes, Eric, *The Life and Masterworks of J. M. W. Turner*, New York: Parkstone Press, 2008.

———, ed., *Turner: The Great Watercolours*, London: Royal Academy of Arts, 2000.

———, *Turner's Human Landscape*, London: Heinemann, 1990.

Shapiro, Alan E., "A Study of the Wave Theory of Light in the 17th Century," *Archive for History of the Exact Sciences* 11, no. 2–3 (1973): 134–266.

———, "Huygens' 'Traite de Lumiere' and Newton's Opticks: Pursuing and Eschewing Hypotheses," *Notes and Records of the Royal Society of London* 43., no 2 (July 1989): 223–47.

———, *The Optical Papers of Isaac Newton, Volume 1: The Optical Lectures 1670–1672*, Cambridge, UK: Cambridge University Press, 1984.

Sharlin, Harold I., *The Making of the Electrical Age*, London, New York, and Toronto: Abelard Schuman, 1963.

Shelley, Percy Bysshe, *The Complete Poetical Works*, Lexicos Publishing, 2012, Kindle edition.

Sims, Michael, *Apollo's Fire: A Journey Through the Extraordinary Wonders of an Ordinary Day*, New York: Penguin, 2007.

Singh, S., *Fundamentals of Optical Engineering*, New Delhi: Discovery Publishing House, 2009 Available at Google Books.

"Slow Light; About Light Speed," Physics Central, website (http://physicscentral.com/explore/action/light.cfm; accessed May 11, 2015).

Smith, A. Mark, *Alhacen on Refraction, Vol. Two*, Philadelphia: American Philosophical Society, 2010.

———, *Ptolemy's Theory of Visual Perception: An English Translation of the Optics with Introduction and Commentary*, Philadelphia: American Philosophical Society, 1996.

———, trans., *Alhacen's Theory of Visual Perception, Vol. One*, Philadelphia: American Philosophical Society, 2001.

———, trans., *Alhacen's Theory of Visual Perception, Vol. Two*, Philadelphia: American Philosophical Society, 2001.

Sontag, Susan, *On Photography*, New York: Picador, 2001.

Sproul, Barbara C., *Primal Myths: Creation Myths Around the World*, New York: HarperCollins, 1991.

Stallings, A. E., trans., *Lucretius—The Nature of Things*, New York: Penguin, 2007, Kindle edition.

Steiner, Rudolf, *Goethe's Conception of the World*, New York: Haskell House Publishers, 1973.

"Stephen Chu: Laser Cooling and Trapping of Atoms," DOE R&D Accomplishments, Research and Development of the U.S. Department of Energy, website (www.osti.gov/accomplishments/chu.html; accessed September 30, 2014).

Stevenson, Robert Louis, "A Plea for Gas Lamps," in *The Works of Robert Louis Stevenson*, vol. 6, Philadelphia: John D. Morris & Co., 1906.

Stewart, Susan, *Poetry and the Fate of the Senses*, Chicago: University of Chicago Press, 2002, 291.

Suh, H. Anna, ed., *Leonardo's Notebooks*, New York: Black Dog & Leventhal Publishers, 2005.

Swenson, Lloyd S., Jr., *The Ethereal Aether: A History of the Michelson-Morley-Miller Aether-Drift Experiments, 1880–1930*, Austin and London: University of Texas Press, 1972.

Sykes, Christopher, ed., *No Ordinary Genius: The Illustrated Richard Feynman*, New York: W. W. Norton, 1994.

Teresi, Dick, *Lost Discoveries: The Ancient Roots of Modern Science from the Babylonian to the Maya*, New York: Simon & Schuster, 2002.

Thoreau, Henry David, *Walden, or Life in the Woods*, New York: Library of America, 1985.

Tobin, William, *The Life and Science of Leon Foucault: The Man Who Proved the Earth Rotates*, Cambridge, UK: Cambridge University Press, 2003.

Tolstoy, Ivan, *James Clerk Maxwell: A Biography*, Edinburgh: Canongate Publishing, 1981.

Tomkins, Calvin, "Profiles: Flying into the Light," *New Yorker*, January 13, 2003 (www.newyorker.com/magazine/2003/01/13/flying-into-the-light; accessed September 25, 2014).

Toomer, G. J., trans., *Diocles: On Burning Mirrors—The Arabic Translation of the Lost Greek Original*, Berlin, Heidelberg, and New York: Springer-Verlag, 1976.

Townes, Charles H., *How the Laser Happened: Adventures of a Scientist*, Oxford, UK, and New York: Oxford University Press, 1999.

Turnbull, H. W., ed., *The Correspondence of Isaac Newton, Volume I, 1661–1675*, Cambridge, UK: Cambridge University Press, 1960.

Turner, Howard R., *Science in Medieval Islam: An Illustrated Edition*, Austin: University of Texas Press, 1995.

U.S. Patent 208,252 A, "Improvement in Electric Lamps" (www.google.com/patents/US208252; accessed May 19, 2014).

U.S. Patent 223,898, Thomas Edison's Incandescent Lamp (http://americanhistory.si.edu/lighting/history/patents/ed_inc.htm; accessed May 19, 2014).

Valarino, Evelyn Elsaesser, *On the Other Side of Life: Exploring the Phenomenon of the Near-Death Experience*, New York and London: Insight Books, 1997.

Valkenberg, Pim, *Sharing Lights on the Way to God: Muslim-Christian Dialogue and Theology in the Context of Abrahamic Partnership*, Amsterdam and New York: Rodopi, 2006.

Van de Wetering, Ernst, *Rembrandt: The Painter at Work*, rev. ed., Berkeley: University of California Press, 2009.

Van Helden, Albert, trans., *Galileo Galilei, Sidereus Nuncius, or The Starry Messenger*, Chicago and London: University of Chicago Press, 1989.

———, *The Invention of the Telescope*, Philadelphia: American Philosophical Society, 1977.

von Franz, Marie, *Creation Myths*, Zurich, Switzerland: Spring Publications, 1972.

von Simson, Otto, *The Gothic Cathedral: Origins of Gothic Architecture and the Medieval Concept of Order*, New York: Pantheon Books, 1956.

Wallace, Robert, *The World of Rembrandt 1606–1669*, New York: Time-Life Books, 1968.

Walshe, Maurice, trans., *The Long Discourses of the Buddha: A Translation of the Digha Nikāya*, Boston: Wisdom Publications, 1995.

Wardrop, David, ed., *Arabic Treasures of the British Library: From Alexandria to Baghdad and Beyond*, London: Friends of the Alexandria Library in Association with the British Library, 2003.

Wechsler, Lawrence, "L. A. Glows," *New Yorker*, February 23, 1998 (www.newyorker.com/magazine/1998/02/23/l-a-glows; accessed September 25, 2014).

Weiss, Richard J., *A Brief History of Light and Those That Lit the Way*, Singapore: World Scientific Publishing Company, 1996.

Westfall, Richard S., *Never at Rest: A Biography of Isaac Newton*, Cambridge, UK: Cambridge University Press, 1983.

White, Michael, *Isaac Newton: The Last Sorcerer*, Reading, MA: Addison-Wesley, 1997.

Widengren, Geo, *Mani and Manichaeism*, London: Weidenfeld & Nicolson, 1961.

Wilk, Stephen R., *How the Ray Gun Got Its Zap: Odd Excursions into Optics*, Oxford, UK: Oxford University Press, 2013.

Willach, Rolf, *The Long Invention of the Telescope*, Philadelphia: American Philosophical Society, 2008.

Williams, John R., *The Life of Goethe: A Critical Biography*, Oxford, UK: Blackwell, 1998.

Williams, L. Pearce, *Michael Faraday: A Biography*, New York: Da Capo Press, 1965.

Wogan, Tim, "Controlling Ferro-Magnetic Domains Using Light," *IOP Physics World*, online journal, August 21, 2014 (http://physicsworld.com/cws/article/news/2014/aug/21/controlling-ferromagnetic-domains-using-light; accessed May 12, 2015).

Wootton, David, *Galileo: Watcher of the Skies*, New Haven and London: Yale University Press, 2010.

Wordsworth, William, *The Complete Poetical Works*, Lexicos Publishing, 2012, Kindle edition.

Wrege, Charles D., and Ronald G. Greenwood, "William E. Sawyer and the Rise and Fall of America's First Incandescent Light Company," *Business and Economic History*, 2nd series, 13 (1984): 31–48.

Wright, M. R., ed. *Empedocles: The Extant Fragments*, New Haven and London: Yale University Press, 1981.

Yonge, C. D., trans., *Diogenes Laertius: The Lives and Opinions of Eminent Philosophers*, Oxford, UK: Acheron Press, 2012, Kindle edition.

Young, Thomas, "Classics of Science: Young on the Theory of Light," *Science Newsletter* 16, no. 447 (November 2, 1929): 273–75.

Zaehner, R. C., *Hindu Scriptures*, London: J. M. Dent, 1966.

Zajonc, Arthur, *Catching the Light: The Entwined History of Light and Mind*, New York: Bantam Books, 1993.

Zaleski, Philip, and Carol Zaleski, *Prayer: A History*, Boston and New York: Houghton Mifflin, 2005.

Index

A Note on the Author

Bruce Watson is the author of four books: *Bread and Roses, Sacco and Vanzetti* (nominated for an Edgar Award), and *Freedom Summer*. While writing for *Smithsonian* magazine, he covered topics ranging from eels to pi to profiles of artists and writers. His work has also appeared in the *Wall Street Journal*, the *Washington Post*, *Los Angeles Times*, and other publications. He lives in western Massachusetts.